MATERIALITIES IN ANTHROPOLOGY AND ARCHAEOLOGY

PLANTS MATTER

MATERIALITIES IN ANTHROPOLOGY AND ARCHAEOLOGY

SERIES EDITORS

Luci Attala and Louise Steel
University of Wales Trinity Saint David

SERIES EDITORIAL BOARD

Dr Nicole Boivin
Director of the Max Planck Institute for the Science of Human History

Professor Samantha Hurn
University of Exeter

Dr Oliver Harris
University of Leicester

Professor David Howes
Concordia Centre for Interdisciplinary Studies in Society and Culture

Dr Elizabeth Rahman
University of Oxford

MATERIALITIES IN ANTHROPOLOGY AND ARCHAEOLOGY

PLANTS MATTER

EXPLORING THE BECOMINGS OF PLANTS AND PEOPLE

Edited by
LUCI ATTALA
and LOUISE STEEL

UNIVERSITY OF WALES PRESS
2023

© The Contributors, 2023

All rights reserved. No part of this book may be reproduced in any material form (including photocopying or storing it in any medium by electronic means and whether or not transiently or incidentally to some other use of this publication) without the written permission of the copyright owner except in accordance with the provisions of the Copyright, Designs and Patents Act 1988. Applications for the copyright owner's written permission to reproduce any part of this publication should be addressed to the University of Wales Press, University Registry, King Edward VII Avenue, Cardiff CF10 3NS.

www.uwp.co.uk

British Library Cataloguing-in-Publication Data
A catalogue record for this book is available from the British Library.

ISBN   978-1-83772-048-4
eISBN   978-1-83772-049-1

The rights of The Contributors to be identified as authors of this work have been asserted in accordance with sections 77 and 79 of the Copyright, Designs and Patents Act 1988.

Typeset by Marie Doherty
Printed by CPI Antony Rowe, Melksham

CONTENTS

	List of Figures	vii
	List of Tables	viii
	Acknowledgements	ix
	List of Contributors	xiii
	Preface	xvii
1	Introduction: Talking of (and with) (the Materiality of) Plants Luci Attala and Louise Steel	1
2	The Materiality of Plants: Plant–People Entanglements Marijke Van der Veen	37
3	Plants as Medicine in the Anthropocene Sarah E. Edwards	57
4	The World Tree: Humans, Trees and Creation on the Sierra Nevada de Santa Marta Falk Parra Witte	83
5	Composing with Plants: Discerning their Call Julie Laplante and Kañaa	109
6	The Matter of Knowing Plant Medicine as Ecology: From Vegetal Philosophy to Tea Tasting in the Anthropocene Guy Waddell	135
7	Escaping to the Garden and Tasting Life Sarah Page	161

8 'The crop that ruled our lives': Memories of Tobacco   189
   among Former Growers in Australia
   Andrew Russell

   Index   215

LIST OF FIGURES

Fig. 4.1　The structure of the cosmos as a World-Tree　89
(Kagkbusánkua) that also ramifies from the centre
of the Sierra Nevada. River valleys and their
corresponding human lineages are seen as branches.
After Parra Witte (2018). Map: Wikimedia Commons.

Fig. 4.2　Seizhua, Mama Shibulata's *eisuama* (top left), partly　97
accepted trees. Before the *nuhué*, stands a sacred
coca-leaf tree. Mama Salé, who grows facial hair,
hails from the forested Hukumeizhi valley (top right).
Mama Bernardo (bottom right) is from the *eisuama*
of Kuamaka (bottom left), which accordingly received
no trees (houses are on the small plateau).
Photos: Falk Parra Witte and Bernabé Zarabata.

Fig. 4.3　Tall 'non-Kogi trees' at about 200m above sea level　99
(top) stand in contrast to smaller 'Kogi trees'
surrounding a traditional homestead at an altitude of
approx. 2.400m (bottom). The imposing tree on the
top left is a Kagksouggi tree on a sacred site, where a
group of Mama are divining. Photos: Falk Parra Witte.

Fig. 5.1　VIBRATION; RESONANCE; TIME　119

Fig. 7.1　'Spending time in the garden completely absorbs　162
and fulfils me' (Peggy).

Fig. 7.2　Pottering in the garden.　164

Fig. 7.3　The unsurpassed taste of home-grown produce.　173

Fig. 7.4　Harvesting salads and foraging for wild garlic.　176

Fig. 7.5　My Initial Vision of the Embodiments of Gardening　178
– The Home-Grown Food Vortex, as the gardener
and eater. Author's own, 2021.

LIST OF TABLES

Table 4.1 The eisuamas' different reactions to being offered trees, in approximate order that they appear in the storyline. After Parra Witte 2018.   93

Table 4.2 Examples of Kogi and non-Kogi trees. Their botanical names were not identified. After Parra Witte 2018.   99

ACKNOWLEDGEMENTS

It is obvious that without plants this book would not have been possible, and therefore, we must acknowledge the part they have played in bringing this volume to press. It is only when one starts paying careful attention to what plants are up to that one begins to grasp the enormity of their process. Plants do not simply provide the stage and the props with which lives are performed, they are the substances that form our bodies and the fabric that holds our lives together, and for that we must all be grateful. However, even knowing this, it is just too easy to overlook what plants contribute. We literally trample over them and assume they will always bounce back up again. And, as we sit here today, during the sixth mass extinction event, cognisant of the accelerating decline of plant (and other) species, we are aware that simply acknowledging plant friends might ring hollow. Nevertheless, is it to the greenery, the vegetation, the undergrowth, the seeds and the weeds that we extend our thanks and would like to recognise a kinship.

Luci's first friendships with plants started at primary school when each child was given a small patch of ground to care for in an area surrounded by the school's orchards. Luci's tiny plot had little on it. Other than a small rose bush and the bumps of a dormant rhubarb plant, it was bare soil. With almost no instruction, the children were left to do as they pleased with their patch. Terrified that her actions could hurt – maybe kill – or possibly anger the plants in some way, Luci tentatively 'gardened'. She pruned her rose bush by cutting off the fat, orange hips, and sat by the rhubarb crowns, not knowing if they were dead or what to do to encourage them to change. Most of that gardening time was spent climbing the apple trees and watching small bugs trundling across the bumpy sods of soil. However, when the spring came, and the plants began to show growth, that time of watching and not knowing was over. Stems poked and stretched, leaves formed ready to uncurl, and buds grew plump preparing to burst. That initial insecurity transformed into wonder and from simply waiting and watching an enduring connection and bond between Luci and that tiny patch was established. Consequently, Luci would like to acknowledge the rhubarb patch, the gnarly apples trees and the little angry pink roses that together formed the core botanical team that began her journey into consciously attending to plants.

That initial team later transformed into the discussions that excited the people who were attending the Botanical Ontologies conference at the University of Oxford in 2014, thus providing the academic soil from which more formalised scholarship could grow. As a result, we would also like to acknowledge the academics who continued to correspond and we have worked with over the years – particularly to Elizabeth Rahman and Guy Waddell, but also Laura Rival and Jeremy Narby. Each in your own way, you have given enormously to our journeys ... perhaps, without realising it, just like the rhubarb, apple trees and roses.

Credit is also due to the botanists whose work has been nothing short of inspirational in recent years – specifically Suzanne Simard, Monica Gagliano and Stefano Mancuso – who have been brave enough to forge new paths for understanding plant-human relationships.

Luci would also like to acknowledge her children, particularly Al whose fierce honesty is both brave and enormously helpful. And finally, to the increasing brood of grandchildren who now run round her garden getting excited about what delights it brings.

Louise's adventures with plants began in her grandmother's garden in Teignmouth. She has particularly fond memories of watching Grandma potting plants in the greenhouse and more so venturing into the garden to gather fresh salads for lunch! All the vegetables that Grandma and Grandpa ate, and many of the fruits, were grown in the garden and Louise would like to dedicate this book in part to their memory. Louise would also like to thank Steve Thomas not only for his ongoing support of her research, but also for reigniting her passion for fresh homegrown produce, and even more so for introducing her to the fascinating world of plants on our doorstep in west Wales. He is a constant source of knowledge on the foods we can forage in the hedgerows, as well as a wealth of information about the wildflowers and trees that grow around us. Travelling through the landscape with Steve has opened Louise's eyes to the delights of this natural and abundant world. Steve also introduced Louise to the *Victorian Kitchen Garden*, and Louise is likewise grateful to the wonderful Harry Dodson and Peter Thoday, who both did so much to rescue and preserve knowledge about plants and gardening lore and more importantly make it accessible and enjoyable. Particular thanks are due to Louise's parents John and Dorothy, whose constant support and encouragement allowed her to explore the world and to write about it!

And of course, we must give our heartfelt thanks to Sarah Lewis who commissioned the *Materialities in Anthropology and Archaeology* series for the University of Wales Press – we greatly appreciate her enthusiasm and help in establishing this series and for her patience and guidance through preparation of the text.

LIST OF CONTRIBUTORS

Luci Attala is the Director of UNESCO-BRIDGES (UK) and a Senior Lecturer in Anthropology at University of Wales Trinity Saint David. Her research focuses on materialities with specific attention to coalescing themes of incorporation, ingestion and the becomings of eco-logical interacting bodies. Luci has completed fieldwork in rural Kenya, Spain and Wales where she has mapped the flows of water through bodies including those of plants and people. Luci is a board member of the Tairona Heritage Trust and is currently occupied with a research project called *Múnekañ Masha: Revitalising Water in the Sierra Nevada de Santa Marta, Colombia* with the Kogi. She works with the Educere Alliance – a global community of educators, scholars and project leaders seeking to innovate in education – that runs out of the University of Oxford, and is a member of the UNESCO-BRIDGES-MOST coalition concerned with championing sustainability science. Together with Louise Steel, Luci is the *Materialities in Anthropology and Archaeology* series editor for the University of Wales Press.

Sarah E. Edwards FLS is Plant Records Officer at the University of Oxford Botanic Garden and Arboretum, and teaches Ethnobiology and Biological Conservation at the Institute of Human Sciences, University of Oxford. She is also an Honorary Research Fellow at UCL School of Pharmacy and a board member of the British Herbal Medicine Association. Her research interests include understanding sociocultural aspects of medicinal plant use within different societies and biocultural diversity conservation in northern Australia. Her latest work collaborating with farmers and artists in south Wales has focused on using a multispecies ethnographic approach to re-evaluate human-plant interrelationships.

Kañaa is a healer and founder of the *Association of Research in the Anthropology of traditional Medicine* (ARAM), https://arametra.org/.

He is the author of *Médecine traditionnelle et savoirs thérapeutiques endogènes* (2018), published with l'Harmattan.

Julie Laplante is Professor of Anthropology at the School of Sociological and Anthropological Studies (SSAS) at the University of Ottawa and has been working on human-plant entanglements in healing since the early 1990s; first at the intersections in-between indigenous and bio-scientific/humanitarian plant and molecule-based medicine in the Brazilian Amazon, later with more attention attuned to both ancestral and clinical bodily, visual and sonorous abilities in healing with plants at two edges of the Indian Ocean (South Africa, Java Indonesia) and more recently in Cameroon. Her approach stems from medical anthropology and science and technology studies (STS), moving towards rhizomic phenomenology in anthropology. Julie's publications include an anthropological film *https://www.youtube.com/watch?v=CMRZRw1z2Fw* (2015) connecting with a book chapter *Becoming-Plant* in *https://www.springer.com/gp/book/9783319480862* (2016), an authored book *Healing Roots. Anthropology in Life and Medicine* (2015, 2018). She is also principal co-editor of *Search After Method. Sensing, Moving, and Imagining in Anthropological Fieldwork* (2020) and two special issues in *Anthropologie et sociétés* (Phénoménologies en anthropologie 2016/Devenir-plante. Enlacements et attachements 2020). She is the director of the research group Planthropolab (*https://planthropolab.com/*), supported by the Faculty of Social Sciences at the University of Ottawa.

Sarah Page is an anthropologist, teacher and musician, and a passionate advocate for sustainability in today's current foodways. She has written for *Organic Gardening*, given cooking demonstrations, and organised and run *Healthy Eating Days* in local primary schools. Her research focuses on home-gardening in the UK as food sourcing and how the relationships we develop with foods' coming-into-being give meaning holistically to human health and how this contributes to the discourse on current environmental and climate issues.

Falk Parra Witte is a Colombian-German anthropologist. Falk obtained his PhD from the University of Cambridge based on long-term fieldwork on the Sierra Nevada de Santa Marta. His research seeks better understanding of Kogi ecology as a complex way of being and knowing organised by life-giving cosmological principles. Falk hopes that communication and collaboration between Indigenous and scientific knowledge can encourage sounder ecological dispositions. He is a board member of the Tairona Heritage Trust which seeks to protect the Kogi environment and to offer the Kogi a voice in the wider world. Falk is currently working the Kogi organisation *Gonawindúa Tayrona Organisation* on a project called *Múnekañ Masha: Revitalising Water in the Sierra Nevada de Santa Marta, Colombia* that aims to find synergies between different approaches that attend to the environment.

Andrew Russell is Professor of Anthropology at Durham, UK. Originally educated in the hard sciences at Oxford, he moved to biomedical anthropology upon graduation. After doing fieldwork for his PhD amongst the Yakha, a group in India previously unstudied by anthropologists, he turned his attention to tobacco. Andrew founded the interdisciplinary Smoking Interest Group which forms a collaboration between anthropology and medical humanities. He works closely with the UK Centre for Tobacco Control Studies, as well as FUSE – a centre for Translational Research in Public health and FRESH – the northeast of England's tobacco control office. His work explores the phenomenology of relationships with tobacco as much as where smoking occurs, policies on tobacco and the history of the changing landscape of tobacco.

Louise Steel is Professor of Near Eastern Archaeology at the University of Wales Trinity Saint David, Lampeter. She has worked extensively in Cyprus, focusing her research on the consumption of Late Bronze Age pottery, and has directed excavations at al-Moghraqa (Gaza) and Arediou (Cyprus). Louise is the author of *Cyprus Before History: from the Earliest Settlers to the End of the Bronze Age* (2004) and *Materiality and Consumption in the Bronze Age Mediterranean* (2013), which explores the interaction of objects in peoples' social worlds. More recently she has been exploring the vital materiality of earthy matters and how these have been entangled in peoples' lives over millennia, including a podcast, *Interactions with clay: The creation of settled communities in the Near East* in the Earth and World Series, Camden Art Centre, *https://soundcloud.com/camden-arts-centre/earth-and-world-interactions-with-clay*. Together with Luci Attala, Louise is the *Materialities in Anthropology and Archaeology* series editor for the University of Wales Press and Assistant Director of the UNESCO-BRIDGES Hub at UWTSD, Lampeter.

Marijke Van der Veen is Emeritus Professor of Archaeology at the University of Leicester, UK. An archaeologist, using botanical remains from archaeological excavations as her primary data, her research focuses on the archaeology of human-plant interactions, the meeting of biology and culture. Projects have included agricultural economies (e.g. Iron Age and Roman Britain; prehistoric and Roman North Africa), food supply to Roman quarry sites in Egypt (Mons Claudianus and Mons Porphyrites), the Roman and Islamic spice trades (Myos Hormos/Quseir al-Qadim), and tracing the dispersal of food crops into North Africa and NW Europe between c.AD 1–1500. She is the author of *Crop Husbandry Regimes* (1992) and *Consumption, Trade and Innovation* (2011), as well as editor of *The Exploitation of Plant Resources in Ancient Africa* (1999), and three issues of the journal *World Archaeology*: *Luxury Foods* (2003), *Garden Agriculture* (2005), and *Agricultural Innovation* (2010).

Guy Waddell PhD fHEA MNIMH is a herbalist and educator who lives and works in London. He was a lecturer in herbal medicine in the Department of Life Sciences at the University of Westminster and is currently Director of Studies at Heartwood Education, which provides foundation and professional level courses in herbal medicine. He has authored *The Enchantment of Western Herbal Medicine: Herbalists, Plants and Nonhuman Agency* (2020) as well as edited a (forthcoming) volume, *Plant Medicine: A Collection of the Teachings of Herbalists Christopher Hedley and Non Shaw*. He has a particular interest in how humans and plants know each other and what may come of this.

PREFACE

This book is one of a series that contributes to what is broadly termed the new material turn in the social sciences. The underpinning intention that coheres the numerous interdisciplinary moves that participate and feed into this flourishing body of literature, is to challenge anthropocentricism (Connolly 2013). This series dethrones the human by drawing in materials. Positioned under the broad umbrella heading of the New Materialisms or New Materialities the series aims to draw in the non-human as agent with a view to both recognize and advocate for the other than human entities that prevail and engage in our lives.

In recognition that these terms are somewhat slippery to grasp we have outlined the following distinctions to put clear water between the terms and demonstrate how we are using them.

Distinctions between materiality and matter

The term 'materiality' describes the quality or character of the material a thing is made of. Its material-ness, if you like. On the other hand, the term 'matter' is used to describe physical items that occupy space (mass). Traditional theories of materiality explore how the objects (made of matter (different materials)) shape the lives of people. New Materialities attends to the materials (matter) that objects are made out of and how those materials influence human behaviour.

Materiality and material culture studies have tended to focus their attention on *things* or *objects* (cf. Banerjee and Miller 2008; Miller and Woodward 2010), especially the things that people make. Scholarship has been less concerned with how materials behave in favour of looking at how people use materials. Materiality studies, therefore, demonstrate a connection between humanity and the things they make and use. In other words, it explores how items reflect their makers and owners and therefore embody meanings.

The New Materialities turn moves away from objects and attends to the materials that the objects are fashioned out of. Turning attention to the materials allows a new dimension to open up whereby the substance a thing is made out of becomes significant. Bringing materials to the foreground not only shows that materials are instrumental in providing the character and meaning of an item but also that the materials themselves are determining – even actively responsible – for the final shape and manner by which the finished article can manifest. Thus, how a material behaves predicates how it can be used (see Drazin and Küchler 2015) and in turn how we understand it. This perspective, following Latour (1993), gives materials a type of agency both inherently and whilst in relationship with other materials (see Barad's concept inter-relationality, 2007). Indeed, using this perspective, it is how materials interact or engage that becomes the place of relationship, creativity and attention. Therefore, the NM draws the materials things are made of into focus and by attending to the behaviours and characteristics of those substances, asks the question 'How do the materials (read: substances) that we make things out of, shape our lives?'

References

Banerjee, M. and Miller, D., 2008. *The Sari: Styles, Patterns, History, Techniques*. London: Berg.

Barad, K. 2007. *Meeting the Universe Halfway: Quantum Physics and the Entanglement of Matter and Meaning*. Durham (NC) and London: Duke University Press.

Connolly, W. E., 2013. 'The "New Materialism" and the fragility of things', *Millennium: Journal of International Studies* 41(3), 399–412.

Drazin, A. and Küchler, S. (eds), 2015. *The Social Life of Materials. Studies in Materials and Society*. London: Bloomsbury Publishing.

Latour, B., 1993. *We Have Never Been Modern*. Cambridge (MA): Harvard University Press.

Miller, D. and Woodward, S. (eds), 2010. *Global Denim*. London: Berg.

1 INTRODUCTION
Talking of (and with) (the Materiality of) Plants
Luci Attala and Louise Steel

Materials are ineffable. They cannot be pinned down in terms of established concepts or categories. To describe any material is to pose a riddle, whose answer can be discovered only through observation and engagement with what is there. To know materials, we have to follow them.

(Ingold 2012, 435)

We create the world that we perceive, not because there is no reality outside our heads, … but because we select and edit the reality we see to conform to our beliefs about what sort of world we live in.

(Bateson 1972, vi)

Under an endless rain of cosmic dust, [where] the air is full of pollen, micro-diamonds and jewels from other planets and supernova explosions. People go about their lives surrounded by the unseeable. Knowing that there's so much around us we can't see forever changes our understanding of the world, and by looking at unseen worlds, we recognize that we exist in the living universe, and this new perspective creates wonder and inspires us to become explorers in our own backyards.

(Schwartzberg 2014)

This book adopts a New Materialities approach to explore the methods and manners by which plants and people forge lives together. It is interested in the physicality of relationships and how material abilities inhibit and make possible the way relationships can be enacted. The book is called *Plants Matter* to indicate that the beings currently classified as plants sit at the centre of the text's focus, but it should not be thought of as a book *about* plants in isolation. Nor is it simply about how people use plants or a series of chapters that demonstrate how plant and human lives sometimes collide. Rather, it is concerned with how plants and people become, and have developed, together. And it does this by tracing their material or physical connections.

The title signals that plant life is noteworthy, that plants matter. It is, of course, self-evident that without those first photosynthesising entities emerging approximately 500 million years ago (Morris et al. 2018) the material changes that have enabled and sustained the evolution of the myriad land-living, breathing morphologies may not have been possible. Consequently, it follows that those early florae were undoubtedly instrumental in reorganising the atmospheric gases that created the climate, the soils that formed the diversity of habitats, and provided the bulk of nutrition that life has enjoyed for millions of years. However, recognition of these consequences is not the main lens through which this text attempts to understand how plants matter either. In contrast, it is interested in exploring how the substances or materials that comprise plant bodies and inform their methods and processes affect, inform and shape other bodies.

Therefore, the ideas, examples and accounts in this book are occupied with drawing the materiality of relationships to the foreground and are less interested in understanding items in separation. The New Materialities approach concentrates its attention on interdependencies and how entities are embedded within a tangled system of material mutuality. The approach is interested in the possibilities that relationships generate but it pays more consideration to the brute materiality of relating and acknowledges the processes of becoming that underpin the emergence of forms, following Kohn's anthropology of life (2013, 2015). Thus, the book's overarching message communicates more than merely how 'individualities need or use each other'. It asserts that rather than discrete, bounded packages appearing as individuals, things are more accurately understood as relationships in action – lively, shifting, affective blendings ecologically reliant on each other in intricate and complex ways and that, through being physically porous and materially permeable, entities are constantly in the process of co-creating each other's existence. Consequently, *Plants Matter* provides examples that show how the linguistic and intellectual habits that conceptually separate plants and people from the base fabric of substances misrepresents how things come to be, and it hopes to achieve this by reminding the reader of how materials are shared, shuffled and are (re)formed through the relentless and persistent processes of gathering, mingling and becoming together.

> Everything may be something, ...[but] being something is always on the way to becoming something else.
> (Ingold 2011, 3)

The grammar of matters

The word 'matter' describes both the stuff of life – chemicals, substances, and any 'things' with physical mass – but, as outlined above, it is also used here to bring in the world as a multitude or tangle of relationships that are constantly producing the stuff that the world and its objects or things are made of. Thus, on some occasions, the New Materialities approach is almost an attempt to intellectually penetrate form because, following Barad (2007, 2010), it peers imaginatively into the kaleidoscopic world of mattering to help grasp the burgeoning fecund material connections that incessantly generate the forms one can see and engage with. Therefore, it is not simply concerned with the social and cultural lives of objects or things. It prefers an interdisciplinary stance that unwraps how items (formed from the substances that collect as morphologies) influence and shape each other, and how existence in all its manifestations – lives, forms, species, things, objects, forces – emerges from (and with) the material identities or characters of different matters as they relentlessly collide, cohere, divide and regroup. Therefore, at its foundation, New Materialities thinking holds that bodies (of any kind) should not be understood in separation, nor should they be thought of as bounded – composed of a cohered chemical assemblage – but should be recognised as 'open', as recursive, relational conversations that use a vocabulary and grammar of matter – to produce temporary entities from the flow and flux of substances. This occurs through numerous, easily recognisable, brute physical processes such as, for example: absorption, ingestion, inhalation, assimilation, invagination, photosynthesis, transpiration, respiration, defaecation, decay – depending on body type – but also occurs at multiple microscopic and quantum scales simultaneously. It is this hidden reality that allows bodies to form as they do and therefore must be including in discussions to avoid the thoughtless damage that is regularly wrought upon matter.

Adopting a New Materialities approach

Take a minute to do this activity.

> Recognising that the combination of gases popularly known as air surrounds everything above land is a useful imaginative device to help one realise how everything is exists in relationships of becoming.
>
> Imagine the air around you for a moment. Ask yourself where it stops and starts. Can you sense how it is relating to you? Is it touching you? Can you feel it? Are you aware of what it is doing?
>
> Air is often represented as the space between what exists or what is experienced as there. If life can be represented by a picture drawn on a white piece of paper, air is the white paper between the things one has sketched. Air is known to be there, but it is also simultaneously experienced as not anywhere. Materially speaking, air not only clings to material entities, but it also penetrates them and connects or joins what appear to be standing apart. This material reality demonstrates the lack of emptiness or space between bodies and any other gatherings of matter. In addition, the air, itself full of dancing gases, microscopic entities and dusty detritus shed from bodies (Coard 2019), is also a shifting set of materials that holds or binds things together. Like fish in the oceans, entities on land swim through the air. There is no space, places are never empty, everywhere is full and connected.

The above exercise perhaps illustrates what this approach emphasises. Specifically, that appreciation on the material interactions and influences that bring entities into existence produces a different world, where all parts and all acts of damage are connected. Consequently, this book hopes to inspire thinking that reduces the intellectual distance that has been placed between forms or mass, and to draw things together as endurances. This approach encourages the 'edges', of what appear to be separate entities (existing in distant homeostatic stability as if alongside each other) to blur and blend. This blending amalgamates and moves the way the imagined Newtonian world functioning as a series of mechanically associated separate working parts to a spooky, slippery, Baradian queer coherence, an alien material unity, a shared *universe* (Barad 2003, 2010). It hopes to stay away from what Adamson (2014, 260) calls a 'clichéd' picture of 'universal connectedness' but it does sidle towards it, and it borrows from Ingold's (2012) 'ecology of materiality', to explore how plant and fleshy matters engage in materially formative performances that fundamentally rely on each other to exist (*in the ways that they do*).

The collection of chapters in this book offers a range of interdisciplinary perspectives on how plant and human bodies share tangled life ways. Often using ethnographic and ethnobotanical information, the chapters explore how the behaviours of certain plants have all but enticed, excited and even seduced people to pay attention to them. Consequently, through forging affective relationships plants are shown to be co-authors of people's stories – responsible for fashioning, organising and even co-designing people's worlds. Thus, how people can know the world should be thought of as achieved-*with* plants because they not only populate the landscape, they also alter human physiology in multiple material ways – through gatherings or sensorial conversations using the chemistry of taste, perfume, colour, sound and textures. Thus, as a collection, the chapters show how together plants and people both instrumentally occupy and organise each other's days.

Why use a materialities focus?

Most of us are not used to looking at the world through a New Materialities lens. This is perhaps because we have been 'trained' to see the world as populated with separate entities concerned with surviving rather than a pattern of mutual interdependencies.

Gregory Bateson (1972, 1979) famously spent much of his time occupied with locating the elusive connecting patterns that generate the forms, forces and systems that order life. He was convinced there is a pattern within the chaotic system of materials shifting and morphing into different forms – lobsters, hands, leaves, mushrooms – and he was sure that 'the major problems in the world, are the result of the difference between how nature works and the way people think' (Bateson 2015). A New Materialities focus hopes to foreground that shifting pattern so that it becomes possible to think in those terms, rather than remain using what Taussig (1992) calls the 'Nervous System' methods of understanding that currently dominate the world.

Thoughts are grammatically organised through language and socio-cultural expectations. For example, my (Attala) thoughts –pictured here as straight-line steps walking out along regimented trails – are positioned in an acceptable order on the dry land of this page. My thoughts, as they emerge to be communicated, are discouraged from anything messy or out of the expected order or place (Douglas 1966). I have been schooled to avoid the muddy, squelching uncertainties of other intellectual, cultural, and topographical terrains and to keep to the paths. This book hopes to encourage what might be expressed as a tentative rewilding of thinking, as a call to walk off the beaten track and shift perspectives towards seeing the undivided or dividual (Strathern 1988).

It is abstract, but not too baffling, to grasp that chemical elements – such as carbon, hydrogen or copper (and any other elements on the periodic table, for that matter) can combine, and in doing so create different substances with different properties. In connection, it is equally intelligible to understand that those substances may continue to collaborate with other substances and become things or objects or beings that too can later alter. Consequently, people are aware that through processes of growth and decay the world is created and recreated by a ceaseless, perpetual shuffling of matters. However, the brute reality of a constantly active material chaos is an almost mythical, fantastic world – and is one that still escapes both human senses and full comprehension when we make choices. Our senses appear to insist we live in a world of bounded individualities, which means only a very limited and distanced appreciation of the responsive materiality of the 'subvisible world' (Sagan and Margulis 1988) is perceptible (perhaps Descartes was correct – we cannot trust our senses?). Indeed, we even flinch nervously at the strange strangeness (Morton 2010, 2013, 2016) of the other micro worlds when told of its mysterious beings, powers and properties (Barad 2010). That collective anxiety was obvious to Bateson (1972, 1979). He was concerned that it produced an underlying drive to control what appears to be around us – our bodies, our homes, our towns, our nations – with boundaries both cognitive and physical. For Bateson, this was a fundamentally destructive habit and he believed that a revolution in thinking was required to break through the dominance of approaches that insist on analytically dividing the world into classified and ordered typologies.

Following Bateson's (1979) assertion that people need to think like nature, the New Materialities move calls for thought that claims to represent how the world works to echo the inherent wanderings and manoeuvrings of worldly substances and to redraw how we design and understand our world in accordance with the way the crazy, quantum superpositions or spacetimematterings (Barad 2010) dance mass into existence.

(Re)Presenting plants and people

> Ideas or theories about human nature have a unique place in the sciences. We don't have to worry that the cosmos will be changed by our theories about the cosmos. The planets really don't care what we think or how we theorize about them. But we do have to worry that human nature will be changed by our theories of human nature. Forty years ago, the distinguished anthropologist, Clifford Geertz, said that human beings are the 'unfinished animals'. What he meant by that was that it is only human nature to have a human nature that is very much the product of the society in which people live. *That* human nature is more created than discovered. We 'design' human nature, by designing the institutions within which people live. So we must ask ourselves just what kind of human nature do you want to design?
> (Schwartz 2015, 10)

> The way into the underland is through the riven trunk of an old ash tree.
> (Macfarlane 2020, 3)

Redrawing the present is important at this time in history when the fragile balance of lives-living is reported as in jeopardy (Edwards, this volume; Waddell, this volume). The dominance of representations and methods that continue to look away from the obviously lively, existential mutualisms and interdependences must be replaced with new approaches that underscore and emphasise how lives can only live together. Having produced cascades of damage in a bid to control the way matters can behave, The Age of the Anthropocene (or whatever title you want to give it, see Waddell this volume), signals that it is now time for humanity to think again and find ways to work *with* the world. In addition, rather as multispecies studies have let non-human animals in, this text hopes to let the world of plants in.

In the last twenty years plants have changed dramatically. Previously established as limited, almost automated, photosynthesising organisms, vegetal beings now present as much more complicated, as communicative, receptive entities with a range of unexpected methods and sense-abilities (Gagliano 2018). Accordingly, a small group of botanists maintain they have now demonstrated that plants not only communicate, they are also playful, can remember and therefore learn. This challenging, disrupting picture suggests plants somehow possess the faculties commonly associated with having a brain (Baluska et al. 2009; Baluska and Mancuso 2009; Gagliano et al. 2014; Gagliano 2015; Mancuso 2021; Pollan 2013). As ways of knowing plants shift, so too have understandings of humanity and how lives have been lived in association with plants. Consequently, representations of humanity's relationships with plants are also being redrawn and new depictions are appearing in their place.

Damaging images

The tendency to imagine life prior to the invention of farming (Diamond 1987) as fraught with insecurities and a short and brutish affair persists – as does the perceived need to militate against what is regularly characterised as a hostile universe. That early humans existed as self-centred individuals desperately hoping to locate food in environments full of aggressive creatures similarly occupied is an enduring trope but also more than likely a fallacy. Today there is an abundance of ethnographic literature that conjures up another world entirely – one that is enacted as a series of cross-species kinship relationships (Bird-David 1990, 1992; Devin 2020; Mosko 1987), where individuals gratefully respect and honour the other lives or materialities that cross their paths (see Parra Witte and Laplante, this volume). These ethnographic accounts rely on lived experience rather than speculation and therefore illustrate how preceding representations are probably built from flawed assumptions, and therefore are inaccurate representations possibly more evocative of the mindsets that summoned them up than any other reality.

Consequently, evidence from hunter gatherers today convincingly confirms that our cousins before them have not struggled in the ways suggested but easily procured regular and abundant sustenance by mobilising around what is characterised as a 'giving environment' (Bird-David 1990). Moreover, notwithstanding any fluctuations in meteorological conditions, evidence shows that days have not been spent hunting or gathering hopefully because the expert knowledge regarding the predilections, locations and seasonal activities of nutritious and medicinal plants passed down to generations have acted to link human bodies to plant bodies from the outset (see Van de Veen and Edwards, this volume). Studies show that gathered food represents up to 80 per cent of hunter gatherer consumption patterns (Boserup 1970) and takes only a few hours a week to achieve (Sahlins 1974), leaving days full of time to dance, sing and play (Milton 2000). Thus, we can put the patriarchally inspired stereotypes of 'man the hunter' (Ardrey 1956) furiously learning to kill animals to survive to bed and spread the word that early days and diets engaged with the edibility of plants more regularly and fruitfully than any other edible substances. Drawing the above together foregrounds a rich picture of how plants and people are irrevocably and crucially tied together and have followed each other around, with plants benefiting from the care and support humanity can provide as much as the other way around (Van de Veen, this volume).

Van de Veen's chapter challenges conventional archaeological perspectives of plants to move away from an anthropocentric view that sees plants as passively useful and towards one that is more inclusive and that acknowledges the role plants have played in shaping human lives. In contesting the notion that plants are passive, she adopts a plant's perspective of their relationships with people. By focusing on the materialities of relationships, Van de Veen goes on to demonstrate the ways in which it is obvious how (and that) plants and people have always been inextricably entwined. Using this truism, she unpacks how the people-plant activities such as domestication, farming, diet and so on, that are *assumed* to have been human instigated and beneficial only to people, need to be reconsidered to acknowledge plants' roles in these associations and thus appreciate how plants and people live together.

It is axiomatically obvious that people everywhere have watched the world changing seasonally, have paid attention, noticed, experimented, remembered and have been provided for by, even owe their existence to, local plants. Moreover, in respect to the empirical reality that lives cannot be, and have never, lived alone, recognising that relationships between plants and other substances influence and affect the wider webs of materiality is an important foundational step necessary to begin rethinking the world.

Applying the notion of a 'giving environment' helps thinking to shift away from imaging the landscape as an intrinsically intimidating arena and move towards visualising a set of dependencies, affiliations and linkages within a material field. This is not an exercise in anthropomorphising the landscape. It is a recharacterisation that hopes to help shift the habit of employing competition as the mobilising force and move towards reframing events as inter-reliant reciprocal relationships. This repaints pictures of early humans from besieged fighters to attentive sharers dependent on their relationships with plants, and this, in turn, also allows human-plant relationships to move away from desperately seeking sustenance to a different modality where curiosity and the intimacy of eating plants urged further appreciation (see Page, this volume). One can assume that this process of learning-with was achieved by attending to plants and therefore it opened the door to knowing plants and understanding how their materiality develops and affects human bodies – rather than controlling plants to accumulate surplus in a bid to survive. Collecting and retaining information about plant identities and their capabilities are the origins of what is now called herbal knowledge (Waddell, this volume) and can be positioned as seeking assistance from the multiple subjects or, in some cases, the vegetal teachers rooted into the landscape (Beyer 2010; Narby and Chanchari Pizuri 2021).

Plant teachers

The notion of plants as teachers can be found in countless ethnographies that detail lives around the world (Beyer 2010; Luna 1984;

Narby 1999; Narby and Chanchari Pizuri 2021; Ravalec et al. 2007). Using these accounts, humanity has not been engaged in simply securing plants to use but signals how plants hold information – or are knowledgeable subjects – that provide important life lessons. The plants considered capable of teaching are, therefore, approached for answers to existential questions when their language is understood (Attala 2017, 2019; Beyer 2010; Gagliano 2018; Narby 1999; Ravalec et al. 2007). One way to learn plants' languages is to incorporate their materiality. This could be through ingestion but also could happen via topical absorption, inhalation, or by infusion. Once inside the body the plant is able to communicate physiologically by affecting the chemistry of the body and altering experience (Attala 2017, 2019; see Edwards and Waddell this volume).

Tobacco

According to Narby and Chanchari Pizuri (2021), tobacco is one of the world's most significant and powerful plant teachers. According to Russell (this volume), tobacco is probably the only plant that has been the subject of an international health treaty orchestrated through the World Health Organization's (WHO) Framework Convention on Tobacco Control (FCTC). Using both characterisations, tobacco emerges as a crop that has ruled people's lives. From an allopathic medical perspective tobacco smoke is held to contain 70 different carcinogenic chemicals and therefore is stated to be an extremely dangerous substance capable of causing substantial harm to people's bodies when inhaled regularly. However, the type of tobacco grown in Peru, and inhaled shamanically, is not. Characterised locally as another materiality entirely this type of tobacco is regularly used as a compellingly potent medicine with a catalogue of health benefits (Narby and Chanchari Pizuri 2021). Therefore, despite botanical classifications, the tobacco grown in the Amazon is celebrated as having the power to heal, while the same plant grown and manufactured into cigarettes in other locations injures. Correspondingly, when smoked as industrially processed cigarettes, Indigenous practitioners acknowledge tobacco is hazardous but simultaneously claim that local tobacco provides a pedagogical gateway into other worlds (Narby and Chanchari Pizuri 2021, 16).

In conversation with Narby, Rafael Chanchari Pizuri (a tobacco expert and elder of the Shawi people in Peru), explains the significance of tobacco in Peru. He outlines how the nicotine in tobacco 'allows us to perceive the imperceptible' (2021, 11), and that by seeing the world as it really is, one learns and is taught by the plant. Tobacco is held close and nurtured by shamanic practitioners across the Amazon for this purpose, but this is not tobacco's only ability, Rafael explains that it also cures:

> Tobacco takes away drowsiness, and it cures incompetence and laziness in people and in dogs. Around here, we talk of people who live 'in the easy' – people who make no effort to learn the skills needed to produce the food they eat to stay alive. Tobacco can cure this. It can also cure people who, out of sheer laziness, don't know how to do things. More generally, it serves to transfer power, capacities, and abilities…[it] also helps heal snakebite and the bite of stinging ants. And it strengthens the masculine and feminine hormones. And when added to other anticarcinogenic plants, it potentiates them.
>
> (Rafael, talking to Narby and Chanchari Pizuri 2021, 14)

In contrast to Rafael's characterisation of tobacco, Russell's chapter in this volume fixes its sights on the effects of tobacco's relationship with people out from the Amazon. It too however speaks of tobacco's agency and explains how, from the local confines of the Americas 500 years ago, tobacco has become the internationally available commodity it is today. Grown (or manufactured) in enormous quantities so as to provide the daily (even hourly) intimacy of inhalation for countless people world-wide, this tobacco exerts another powerful hold over the people of the world. Therefore, taking tobacco-the-crop as its starting point, Russell's chapter explores how, through multiple familiar relationships with people, tobacco, once influential in the Amazon, now sits at the centre of a global network dedicated to its existence. Russell focuses attention primarily on the cultivation of tobacco, now achieved at a plantation-style industrial-scale, in Australia and Aotearoa/New Zealand, but he also explores how tobacco continues to generate consequences in the way humanity acts, understands and uses the plant.

The above briefly illustrates that in relationship with human flesh, tobacco sits at an interesting intersection between harm and health. However, in and of itself, tobacco is considered to have negligible influence on people-as-bodies until it is incorporated in some way thus perpetuating a picture of incidental collision and circumstance. Both perspectives of tobacco recognise how the plant has shaped human experience, but they leave the plant dumb and incapable until it penetrates and acts with one's flesh.

The social lives of communicating plants

Echoing the claims of knowledgeable plants made by Indigenous practitioners (e.g. Beyer 2010; Kimmerer 2014; Narby and Chanchari Pizuri 2021), a rapidly expanding body of botanical literature is beginning to demonstrate that plants can do more than previously imagined. Since the early 2000s, findings reached in experimental conditions, are starting to expand conventional definitions and consequently calls for plants to be recognised as responsive beings with a repertoire of surprising skills and proficiencies are being heard (Attala 2017; Gagliano 2018). Indeed, in 2009 Baluska and Mancuso even went so far as to declare plants would be better described as social organisms because of the rapidly building catalogue of results that suggest plants act cooperatively and in ways that imply they are aware of others.

To support Baluska and Mancuso's (2009) call, much of this experimental work exposes plants as expressive communicators able to use a range of both chemical and audible vocabularies to transmit a diverse array of messages across species' boundaries (see Baldwin et al. 2011; Cahill et al. 2010; Dudley and File 2007; Gagliano 2013a, 2013b; Gagliano and Renton 2013; Karban et al. 2006, 2013; Simard 2009a, 2009b, 2012, and an overview of this work in Attala 2017, and Waddell, this volume). These findings not only reveal how, but also makes suggestions as to why, plants communicate. Overwhelmingly these results confirm that plants influence conditions through the release of different volatile organic compounds (VOCs). VOCs are used for a suite of purposes and are reported as ecologically functional because they are deployed to attract, deter and mediate relationships with various nearby entities including plants, insects, and herbivores (Meents and Mithöfer 2020). Consequently, using this broad vocabulary plants send messages that chemically affect, respond to and manipulate the world around them.

Following this literature's conclusions, the puffs of chemicals (VOCs) that fill the air can be warnings, appeals, gifts and as acts of care or aid for what are sometimes termed kin plants. However, because the chemistry of emitting VOCs is enormously complicated the reasons cited for plant emissions continue to raise the eyebrows of some scholars (Alpi et al. 2007; Taiz et al. 2019) who call for caution when depictions and conclusions extend past what a plant is commonly defined as able to do. For example, one might not be surprised to know that emissions are thought to be used for defence purposes (as it fits with the ideas of selfishness now embedded in genetic theory), but one might be puzzled to hear that they are used as support mechanisms that provide benefits for both the emitting plant *and* those being communicating with.

Orthodox understandings have established plants as mute and almost mechanical in their methodologies, splendid but limited in ability, buffeted by conditions and the vicissitudes of circumstance, and because they are unable to move are at the mercy of geographical fluctuations. This is a persuasive, seemingly sensible and empirically obvious perspective but it can also be argued that this view of plants is grounded in the kind of human exceptionalist and patriarchal thinking that has established what constitutes complexity and that certain complexities indicate superiority by using a human template – think: the Vitruvian Man. Accumulating evidence causes tension to these assumptions and suggests that firstly, sophistication can be established otherwise and secondly, that differences do not need to be characterised hierarchically. Indeed, challenging the penchant for hierarchies is an important component in producing equality in the broadest sense. If we are to successfully level the representational playing field, placing differences into hierarchies needs to either become a quaint custom of the past or rethought entirely.

In sum, it is becoming hard to ignore that plants are more than once assumed; they are showing themselves to be aware of their surroundings and that the material consequences of their behaviours action more than individual persistence over others. As life for plants is revealed to be about more than basic survival, the example of trees communicating in forests is a striking illustration of how plants do have something like social lives.

Forests and mother trees

The now widely familiar work of Simard et al. (2011, 2012, made popular now by Simard's book, 2021) provides an astonishingly vivid account of how trees in a forest care for each other. Simard demonstrates how trees work in association with fungi to provision assistance across the forest. This is achieved through the delicate intricacies of their root systems as they entwine with mycorrhizal fungal networks to form a vast web of connections that shuffle nutrients from areas of excess to those that are in need. Simard (2021) paints a picture of trees previously unimagined – one of cross species mutualisms where entities are cooperatively engaged in directing supplies to each other. But this isn't all she has done. Her work has introduced the notion of mother trees. Mother trees, she explains, are pivotal to forest survival. They are the more established trees and they care for the younger members of the forest by reaching out and remaining connected to them below ground. This information has helped reassess representations of trees as separate entities trying to survive alone. In disagreement with the notion that life is comprised of individuals focused on their genetic survival, this work (echoing Bird-David's 1990 notion of the 'giving environment') again elucidates how lives work together, across species and, moreover, by processes of giving and sharing, not storing and accumulating. Thus, trees not only sense 'need' in others, but they also act to remedy that need by redistributing resources. Simard (2021) appears to have no problem with describing trees as social.

Parra Witte's chapter in this volume extends the notion of mother trees used by Simard (2021) out from the forest and into a spiritual landscape of origins. Parra Witte uses the information he collected in conversations with Kogi leaders whilst doing fieldwork in Colombia. The Kogi, an Indigenous group living in the Sierra Nevada de Santa Marta, provide a rich and detailed explanation of the significance of trees in a world that they say is held together by a weave of mystical threads. The use of the word 'mother' signals origins and creativity in Kogi cosmology and the term is used regularly to denote the original creative forces that are responsible for the network of ecological connections that tie people to places, to correct practice, to health, to the past and to the creative spirit of the world, named *Aluna*. For the Kogi, the first or mother tree functions as an axis or pillar around which the world could be formed. Demonstrating some similarity to Simard's botanical findings that position mother trees as central providers, the Kogi use the notion of mother as the first or original tree. In this case, the first plant from which everything else grew. Parra Witte explains that today all plants remain connected to that original mother plant 'because she was the first trunk, from which all had come… in sum, the entire universe, everything that forms part of it […] has its place in this immense genealogical tree' (Reichel Dolmatoff 1985a, 155, 156; 1985b, 86; Parra Witte this volume).

The symbol of a cosmic tree is a regularly recurring motif used to visualise the connections between areas of the universe. For the Kogi, the tree is not a metaphor nor are trees simply symbolically useful. Physical trees, rooted into certain spiritually significant locations, are the

> Key agents in an ontological order that defines trees and humans (and other natural elements) in corresponding, interrelated ways. Beyond making a good model for human genealogical connections (Rival 1993, 11) then, trees are a principal embodiment or reproduction of this order of life
>
> (Parra Witte this volume, 86)

Plant relatives and families

Claims that connect humanity to plant life appear alien at first glance and certainly raise questions about how to understand these associations. Haraway (2016) bravely asks us to make kin with all the 'critters' at this time of trouble but acknowledges that appreciating we are all related (both by being in relationship and/or connected by a shared materiality) is not easy for minds trained to think in terms of separation and boundaries. However, she believes it is an important first step towards creating an eco-just future. Haraway (2016) reminds us that ancestry or genealogy and what might be thought of as natural ties have already been significantly unravelled by feminist and posthuman studies, which has allowed the notion of kin to expand outwards to receive 'others' into the fold. Taking the baseline of a common materiality as our starting point, it is possible to extending our thinking into what, if anything, kin might mean for plants.

The phrase 'plant kin' appears regularly in botanical literature. It does not refer to how humans recognise plants of the same species family but is used to describe plants that have spawned from the same mother plant (Biedrzycki and Bais 2010). In botanical terms, the notion of 'kin recognition' is used to describe the ability of plants to distinguish their own relatives from strangers. It is not clear how plants recognise their kin, or to what extent this is possible, but there are studies that show kin plants share and support each other in ways that non-kin do not (Biedrzycki and Bais 2010). As with the example of trees above, this has been primarily evidenced by underground activities such as root growth (see Karban et al. 2006, 2013). However, some studies also now provide evidence that kin recognition is apparent by some aboveground behaviours. For example, Crepy and Casal's (2015) study shows that kin plants of *Arabidopsis thaliana* orientate their leaf growth to avoid shading relatives and thus promote light sharing between kin, but do not behave in this way with so-called stranger plants.

Establishing that plants recognise kin makes it easy to imagine that cooperative behaviours are restricted to genetically related organisms, but this is not the case. Simard's (2021) work demonstrates that kinship ties are not limited to intraspecies siblings and parents but also cross species boundaries in forests. To imagine that kinship ties are limited to consanguine or genetic similarity relatedness also overlooks the many cross-species relationships that people around the world are familiar with. Indeed, the multispecies ethnographic literature has done a good job in demonstrating the material inaccuracies of imagining bodies can exist without vast populations of other inhabitants (see Gilbert 2017). Furthermore, and bringing it back to the conventional notion of kin, recognising non-human animals as kin does not receive much ridicule or even scepticism these days (Hurn 2012). Indeed, it has become commonplace to describe pets as children and treat them as integral to the family (Haraway 2003; Kohn 2007), even if some individuals find this problematic (Nast 2006; Tuan 1984). On the other hand, recognising plants as part of one's family might be harder to accept. Plants, according to establishment definitions are fundamentally different to animals – another kingdom entirely – and any such cross over must, by these definitions, be absurd. However, claiming a plant or some plants are one's kin is not unusual or contentious in certain parts of the world.

> 'I am *not* dancing alone,' he said. 'I am dancing with the forest.'
> (Kenge talking to Turnbull 1994, 272)

Both the Baka and the Mbuti hunter gatherers who live with the last remaining patches of tropical rainforests in central Africa describe the forest they dwell with as family. The forest is interchangeably described as a father, mother, lover, friend (Mosko 1987; Turnbull 1994) because the Baka and the Mbuti realise (make real) that the forest births the beings that depend on it. Mosko (1987) explains how the forest is the overarching ecological organising principle – it is simultaneously characterised as nurturer, provider and ultimate authority. It is also an animated, emotional material entity that can create conflict, withhold, be entertaining and amusing like any other agent might be. This multi-faceted conception of an assemblage of matters produces an ecological picture of multiple subjects living as layers upon layer of materiality woven together, living within and *as* an African forest simultaneously. Using this perspective, the influencing intersections and associations between different types of flesh-living make the classifications used by other cultures inadequate, unable to describe how the forest lives. Therefore, using this perspective, life living as a forest cannot be divided into typologies successfully and might be better explained as it is experienced by the Mbuti and the Baka – as a material unity constantly in process or becoming (Haraway 2016).

> It holds you close, it protects you. Slithering over your feet, flying into you, scratching your skin. Swimming around you when you step into water. Wherever you are you hear its voice. Its smell. Its breath. Even if you wanted to, you cannot hide, because it has eyes everywhere. Little by little it transforms you. It swallows you up and begins to absorb you – it digests you. The forest. In the end you are part of its body, like the antelopes and the streams. The plants and the palm grubs. You have become an extension of it. You belong to the forest.
>
> (Devin 2020)

Joining plants to people. Do plants communicate with people?

From the above one can get a glimpse into a world populated by a coherence of lively subjectivities, where any firm borders that separate entities are not used to understand existence and where entities experience, or sense, connections physically. In these worlds, plants and others influence each other and both communicate and share in multiple ways. In these worlds, plants and people cohere. In other worlds, where different stories are told, plants are characterised as unable to communicate with people. Here, the notion that plants are aware of humanity is thought of as fanciful nonsense.

Regardless of any incredulity from the establishment about plants' abilities, the work cited earlier has revealed capacities previously unthinkable and in consequence, has prompted further experimentation to determine a fuller understanding of what plants can do. Nevertheless, and despite the careful use of conventional methodologies and the raft of publications in reputable journals, many hard science authors remain unconvinced and are troubled by the use of language that they claim anthropomorphises plant behaviour (Alpi et al. 2007; Taiz et al. 2019), This literature, they maintain, has pushed too far and, as a result they are certainly not prepared to consider if plants communicate with people, other than incidentally. It remains only the wacky few who are prepared to approach, let alone attempt to answer, the question 'do plants communicate with people?' (see Gagliano 2018).

As already outlined, there is a burgeoning body of evidence that shows plants call to hordes of beings – herbivores, parasites, insects, frugivorous animals and so on – yet it continues to remain comical to imagine that they might be communicating with people. Indeed, the text titled *Plant-Animal Communication* by Schaefer and Ruxton (2011) expressly explores how plants communicate with animals – elephants, hummingbirds, bats, sheep – but fails to include people in that category other than noting briefly, and in passing, that humans might disperse seeds with their shoes. Consequently, these authors assume that humanity is excluded, seemingly exempt from plant influences in the way other entities are not. Therefore, human animals are disregarded as either emission beneficiaries or as being on plants 'radars'. Plants it seems couldn't possibly be able to, or be interested in, influencing people from their lowly position in the taxonomic scale. A conclusion that sidesteps the brute materiality of relationships and the overwhelming evidence that human animals are nevertheless continuously influenced by plant life (see Attala 2017, 2019; Marder 2014; Pollan 2002).

> Which raises the question: in the light of evidence that indicates plants are regularly busy persuading all sorts of different species to support their lives using their chemical vocabulary is it likely that humans are exempt from their charms and are not beneficiaries of their instructions? Indeed, one could further ask – why do researchers regularly sidestep this issue?
> (Attala 2014, 2).

This representational blind-spot confirms how some people position themselves in association with plants and reveals how ideas about people and plants continue to be framed by the assumptions embedded into a human exceptionalist mindset – a mindset that stubbornly conceals what is analytically obvious. Plants act upon and affect human animals in multiple diverse ways such as: scent, colour, thorns – let alone through the effects they trigger in bodies through ingestion (see Attala's Edibility Approach 2017, 2019).

Regardless of any of the contentions that ask scholarship that arrives at these conclusions to calm down and step away from this area (Alpi et al. 2007; Taiz et al. 2019), both camps can surely agree that it is patently obvious that:

1. plants react to (or sense) their internal condition,
2. plants are aware of the condition of others and their surroundings, and
3. act in ways that anticipate and pointedly influence other life forms.

In the light of the above, it is clearly time for any notion of plants being mute and distant to be put to bed.

Healing and composing: Extending plants' communicative abilities into bodies

In hoping to reframe plants, Laplante's chapter opens by questioning the value and by considering the consequences of using Aristotle's hierarchical classification of beings that consign plants to the lower levels of existence, as resources for use. Despite being known as the father of science, Aristotle's rationale for organising life into a hierarchy – rather than just groups – is arbitrary, perhaps associated more with Christian thinking than species abilities as the model implies. Plants are placed near the bottom because they are not like animals, she explains; they are stationary, and they do not have obvious sense organs. Creating bounded categories is a recurring obsession that organises how the world is perceived. However, it has its flaws. It doesn't take long to bring up category breakers – think: oyster, for example... is an oyster an animal? Think Mimosa[1] or the Venus fly trap. Are these moving and sensing plants better described as animals?

Critical of the type of linear thinking that created and established these kinds of hierarchies, Laplante connects the 'enslavement' of vegetal lives to the social or 'nervous system' (Taussig 1992) of modern society that is unable to think in 'rhizomic' or 'non-linear' ways. For Laplante, agricultural processes place plants in the service of humanity because they ignore the vitality of these beings, their many abilities and their methods of influence, and rely on imagining people and plants exist separately. Plants, she shows, are not incidental to others' lives and, despite representations, can be (and often are) experienced as agents who sit 'between the living and the non-living' (Laplante, this volume). In addition, Laplante elucidates the connections – seemingly invisible to many – between vegetal and other lives through exploration of how various plants call to, and act on, humanity. By using ethnographic experiences collected with shamanic healers around the world, Laplante argues for alternative methods that recognise and grasp how plants employ 'aromas, dreams, sonority and duration' to affect people (Laplante, this volume, 110).

Recognising that we are at a point in history when plants need to be brought in seriously if we are to negotiate better and fairer ways to live, Waddell's chapter uses embryology and drinking tea as springboards to think through how the similarities and differences between plant and people's anatomy affect shared experiences and ways of knowing each other. Waddell, thus, develops a vegetal philosophy and a vocabulary to encourage people to know plants differently, and, in consequence, to inspire changes in the way plants and people live together. Advocating for 'plantabilities' and cultivating 'plantfulness' rather than mindfulness, Waddell considers how the condition of being rooted may provide a kind of attention and awareness of place that being animal with a mind may not easily access. Waddell's chapter also encourages Attala's (2017) Edibility Approach be stretched to include a 'medibility approach' offshoot – one that recognises the many ways that people and medicinal plants converse with each other through ingestion and assimilation. Using the notion of blending Materialities, Waddell wonders if a focus away from separations lies at the core of producing the wellness needed to negotiate the turbulence of the Anthropocene.

In light of the compounding health and environmental issues of the day, both Waddell's and Edwards's chapters think through the theme of plants as medicine in the Anthropocene. Edwards begins her discussion by noting the many ways humans are not only fundamentally dependent on plants, but also should not overlook the significance of this fact. She then turns to outline how cultural meanings associated with what constitutes a remedy or a treatment contribute to shaping the way that medicinal relationships between plants and people can be approached, articulated and enacted. Moreover, by drawing on her personal experience of working with Euro-American hard scientists, Indigenous/ First Nation and small-scale farmers in Wales, Edwards demonstrates how profoundly different ontological positions continue to cause friction with regards the meaning, ownership and use of plants as medicines either herbal or pharmaceutical.

Page's chapter continues the theme of health and plants with an exploration of how English people turned to gardening during the Covid-19 pandemic 'stay-at-home' order issued in 2020–1. By using gardeners' voices throughout, Page provides a colourful snapshot of the meanings associated with being-in-the-garden when mobility was legally restricted. Research into gardening tends to focus on the health and wellbeing benefits that being active out of the house provides, but Page sidesteps this literature to focus her attention to the value of the extremely British notion of 'pottering' in the garden daily.

The Potawatomi scientist Kimmerer uses the notion of 'species loneliness' to explain a disconnection between humanity and the rest:

> Species loneliness—this deep, unnamed sadness—is the cost of estrangement from the rest of creation, from the loss of

> relationship. Our Potawatomi stories tell that a long time ago, when Turtle Island was young, the people and all the plants and animals spoke the same language and conversed freely with one another
>
> (Kimmerer 2014, 21)

Explaining that the phrase escaping to the garden was not simply a welcome distraction from the world inside the home, Page illustrates how her interlocuters described pottering as a method that manages to gently sustain a connection to what Ingold (2010a and 2010b) might describe as life's flux and flows. This is achieved sensorially, through mindless (or plantful, following Waddell, this volume), corporeal contact with entities that move in a different temporality to that of human lives. Pottering, therefore, offers a type of immersion where one finds oneself in an alternative, lively, tasty and nourishing taskscape (Ingold 2017) that communicates non-verbally, through the senses, in multiple shifting ways. Page's chapter demonstrates that the meanings of gardening emerge phenomenologically and sensorially, and that pottering in this way should be included in the many health benefits of gardening.

The above adequately demonstrates that representing plants as sitting immobile, inert, powerless and serendipitously simply surviving in the landscape is erroneous and needs to be remedied. Plants need to be recognised for their charms and powers to entice, enchant, warn and inspire, as much as fight back, restrict, tangle, obstruct and co-create our lives; plants (like all things) are actively rearranging the substances or the materiality of existence. Consequently, understanding plants' routines and capacities rather than mundane necessity, sits at the core of how lives can be lived together well.

Towards a language of relationships: Using language to draw the world of materials, forms and mass together

> How did language come to be more trustworthy than matter? Why are language and culture granted their own agency and historicity while matter is figured as passive and immutable?
>
> (Barad 2003, 801)

> Language is not the roots of single words, but rather the soil of grammar and syntax, where habits of speech and therefore also habits of thought settle and interact over long periods of time. Grammar and syntax exert powerful influence on the proceedings of language and its users. They shape the ways we relate to each other and the living world.
>
> (Macfarlane 2020, 113)

The characterisations of things – in this case, plants and people – simultaneously enable and restrict how they can be understood. Barad wants to know why representation or significations of the world are held in higher esteem than knowing it sensorially and why it is assumed that matter is non-agential and pliable. Here we are victims of the habits of representation that language has encouraged, which essentially *separate and distance* humans from the material world (Govier and Steel 2021, 304–06). Bateson (1979), like Barad (2003, 2007, 2010), has highlighted and explored the limits of language. For Bateson (1979), the problem with language is that it requires thoughts be shaped in accordance with the shared utterances. Thus, it creates profiles (almost draws outlines around, cuts out, and names shapes in the landscape) and, in so doing, contours and structures how one can communicate and perceive, and making a language of relationships tricky.

> Language continually asserts by the syntax of subject and predicate that 'things' somehow 'have' qualities and attributes. A more precise way of talking would insist that the 'things' are produced, are seen as separate from other 'things,' and made 'real' by the internal relations and by their behavior in relationship with other things and the speaker.
>
> (Bateson 1979, 61)

This suggests that the processes of saying and seeing are – rather like ouroboros – cyclic, leaving one unsure if the world is accurately divided into things or if language has made it so. Thus, the 'things in their thingish world' (Bateson 1979, 61) – in this case, the plants in their plantish world are made 'real' (known in the way that they are thought to be) because they are established that way through language. Linguistic relativity (see the Sapir-Whorf Hypothesis, Sapir 1929) maintains that perception and language co-create each other and therefore one's worldview is restricted or affected by whatever terms are available. Following this train of thought, Bateson (1972) was troubled by dint of the fact that despite things manifesting materially, they are understood or apprehended in cognitive rather than empirical terms and therefore exist in our minds culturally without the terminological tools or ability to 'see' the intricacies of relationships adequately.

The idea of 'warm data' proposed by Nora Bateson (2018), Gregory Bateson's daughter, leans towards finding a language to express interdependencies. The concept emerged from concerns that conventional methods require things be taken out of the overarching pattern to understand them and that in forgetting to replace them the method perpetuates the damaging misunderstandings that have allowed many methods to break the pattern. The notion of warm data sits in contrast to cold or big data and attempts to provide an acceptable model or container in which elucidating interdependencies can become a recognised research method or direction.

Ingold (2010a, 2010b), on the other hand, critical of the use of the word data at all, prefers to use the term 'flux' to summon a sense of an everchanging flow of shifting, unsettled materials. Flux refers to a series of changes, often characterised as a flow but also as a state of constant change – like a river of water moving through the landscape.

Heraclitus famously stated that one can never stand in the same river twice – an analogy that is well used and is useful to us here again. While the river appears the same, because the water is flowing, it is constantly shifting and changing and thus one can never re-enter the same body of water. The notion of the world worlding materials in flux allows one to transport Heraclitus' notion out from the river and up on to the banks. Removed from the river that analogy demonstrates that, despite appearances, one can never step into the same world twice, as the incessant worlding is relentlessly in flux. Moreover, as we are part of the air airing, the plants planting, the soil soiling and so on, it may be more useful to remember that bodies or forms swim within and through the flux together temporarily forming what might be usefully represented as jelly edged congealings in the swirling soup of universal materials (pers. comm. Camille Sandal 2022).

The New Materialities approach, therefore, is determined to conceptually blend everything back the fabric of materiality having now seen what happens when stuff is teased apart into the threads of names. New Materialists and some others, recognising the limits of terminology, explore how one can talk of things successfully when the thingyness of entities is predicated on a chaotically shifting shared field of physicality without edges (see Govier and Steel 2021; Ingold 2012). However, by focusing on the chemistry of existence, plants – or any other thing, for that matter – start to dissolve into the field of substances and then re-materialise intellectually as temporarily coagulated dependencies whose existence is predicated on the surrounding materiality (Barad 2010). Using this lens, the word 'plant', then, holds multiple meanings and if one was to push this to its logical extreme, plants would paradoxically vanish, unable to exist *as such*, but also only able to exist in association with a broad sweep of materials working in unison together. Everything therefore appears to exist in a moving materiality that relies on all the other parts to be itself.

In a search for routes to a common world where peace is possible, Latour (2004, 457) is critical of the consequences of the dominance of naturalism. By taking the perspective of someone attached to the ideas of naturalism he provides a quote, that imagines how such a person might articulate their ideas.

> We all live under the same biological and physical laws and have the same fundamental biological, social, and psychological makeup. This, you have not understood because you are prisoners of your superficial worldviews, which are but representations of the reality to which we, through science, have privileged access. But science is not our property; it belongs to mankind [sic] universally! Here, partake—and with us you will be one.
>
> (Latour 2004, 458)

For Latour (2004), it is this perspective that gets in the way of peace; it is not necessary to explain why it is problematic. When one sees phrases such as 'superficial worldviews' used to describe any understandings other than those presented by the hard sciences, one can only respond with a weary sigh. However, to make another point, it is also possible to flip Latour's quote to make the speaker an

Indigenous practitioner, an elder or even a plant teacher like tobacco, hoping to persuade the hard scientist that they too are trapped in their methods and corresponding worldview! And below we offer you an imaginary perspective of a plant speaking to uncouple some of the authority from the establishment.

> We all live under the same biological and physical laws and have the same fundamental biological, social, and psychological makeup. This, you have not understood because you are prisoners of your superficial worldviews which are but representations of the reality to which we, through… [being plants], have privileged access. But … [plantiness] is not our property; it belongs to [materiality] universally! Here, partake – and with us you will be one.
>
> (Adapted from Latour 2004, 458)

Playing with positionality and situation at a time when existence is dramatically changing assumes importance. It is time to draw in other perspectives, in other authorities. As the altered quote above implies, it is a mistake to suggest that plants live in a different world to people. We inhabit and exist as the same materiality and by shuffling and sharing behaviours we learn from each other. And so, perhaps it is time to ask what it means to think with the plants (Marder 2014) and to draw the vegetal in more equally. To ask what kind of world would we inhabit if the concrete pavements of thinking cracked and let plants sprout through? Stern, for example, is inextricably drawn to two tiny tomato plants courageously taking root in the no-man's land on the border between San Diego and Tijuana, one 'that grows out of a crack in the concrete; the second [which] grows a short distance away in a trickle of water beside the dusty unpaved road concrete' (2017, G25). These have seeded in an unforgiving world amidst the detritus of modern living, reminding us of the power and tenacity of plants.

Indigenous knowings: Decolonising plant knowledges

A key thread running through *Plants Matter* is the very real and urgent need to listen to, and indeed engage with, understandings of what plants *are* and what they *do* drawing upon the rich reservoir of *Indigenous knowings* from around the world. Various chapters

(Edwards, LaPlante, Parra Witte) highlight an intellectual paradigm shift away from the empirically-based, positivist classificatory systems that so-called Western knowledge (the sciences in particular) imposes upon the natural world – identifying, recording, cataloguing and typologising as a means of intellectual validation and control – and instead promote an understanding that the worlding world is more complex and blurred in its becoming. They draw attention to the many voices of peoples who recognise the co-becomings of the world and the 'intimate relatedness and connectedness between humans and non-humans entities' (Blair 2019, 208). This chimes with a very real need to decolonise research and instead to develop new inclusive forms of knowledge production which do not appropriate and control (see Edwards, this volume) local knowledges and belief systems (Archibald et al. 2019a, Archibald et al. 2019b; Thambinathan and Kinsella 2021; Tuhiwai Smith 2021).

The importance of listening and attending to local knowledge comes to the fore in Parra Witte's chapter exploring the Kogi's unique and sustaining relationship with trees, while Edwards reminds us of the negativities implicit in colonial appropriation and control of plant knowledges for our own ends, without *listening to* and acknowledging Indigenous ownership and intangible heritages. Blair's lilyology methodology (2019) conceptualises an in-between space where the *brick wall* of academia can co-exist equally with the rhizomatic Indigenous knowings and '[t]hrough co-existence they can create powerful and dynamic dialogues and discourse leading to transformational learnings, teachings, and knowings' (2019, 211). There are no right knowledges or *storys*, but instead this open dialogue recognises the innate and unique possibilities of different ways of knowing the world.

The urgency of attending to these Indigenous knowings is highlighted by Steffensen, who notes how European colonisation of Australia 'systematically destroyed an invaluable knowledge system that has developed over 1000s of years' (2020, 236) whilst also emphasising how this 'Indigenous Knowledge system is a treasure that holds the answer to many of the social and environmental problems *we all face today*' (2019, 236, our emphasis). Indeed, the very real need to listen and learn, whilst respecting and not re-colonising the voices of the world's Indigenous communities underpins the aims of the UNESCO-BRIDGES initiative:

with the aim of bringing their ecological knowledge and practices into the scientific domain, by providing a transdisciplinary space for indigenous tradition to speak in its own terms with modern environmental methods ... [and] drew attention to the meaning of 'ancestral knowledge' of the land and to the significance of 'sacred sites' in maintaining its health.

(UNESCO-BRIDGES 2019)

This, in line with Indigenous storywork, provides an ethical framework for engaging with the rich seams of knowledge embedded in Indigenous traditional lifeways and 'offers hope at a time when environmental and social crises threaten all life and ways of being' (Archibald, Lee-Morgan and de Santolo 2019b, 8). As Blair reminds us, Country (the land) is far more than 'a passive backdrop to human experience, a scene in which humans live their lives, *a place in which to embed academics' research*' (2019, 207, our emphasis). Instead, the land, and all the beings it supports and sustains, are *active participants* shaping and creating the worlding world, in its ongoing matterings (Barad 2003, 2008; Govier and Steel 2021).

References

Adamson, J., 2014. 'Source of life: Avatar, Amazonia, and an ecology of selves', in *Material Ecocriticism*, ed. S. Iovino and S. Opperman. Bloomington and Indianapolis, IN: Indiana University Press, pp. 253–68.

Alpi, A., Amrhein, N., Bertl, A., Blatt, M. R., Blumwald, E., Cervone, F., Dainty, J., de Michelis, M. I., Epstein, E., Glaston, A. W., Goldsmith, M. H. M., Hawes, S., Hell, R., Hetherington, A., Hofte, H., Juergens, G., Leaver, C. J., Moroni, A., Murphy, A., Oparka, K., Perata, P., Quader, H., Rausch, T., Ritzenthaler, C., Rivetta, A., Robinson, D. G., Sanders, D., Scheres, B., Schumacher, K., Sentenac, H., Slayman, C. L., Soave, C., Somerville, C., Taiz, L., Thiel, G. and Wagner, R., 2007. 'Plant neurobiology: No brain, no gain?', *TRENDS in Plant Science* 12(4), 135–6, doi: *10.1016/j.tplants.2007.03.002*.

Archibald, J., Lee-Morgan, J. and de Santolo, J. (eds), 2019a. *Decolonizing Research: Indigenous Storywork as Methodology*. London: Bloomsbury Publishing.

Archibald, J., Lee-Morgan, J. and de Santolo, J., 2019b. 'Introduction. Decolonizing research: Indigenous storywork as methodology', in *Decolonizing Research: Indigenous Storywork as Methodology*, ed. J. Archibald, J. Lee-Morgan and J. de Santolo. London: Bloomsbury Publishing, pp. 1–15.

Ardrey, R., 1956. *The Hunting Hypothesis: A Personal Conclusion Concerning the Evolutionary Nature of Man*. New York, NY: Atheneum.

Attala, L., 2014. 'Conversations over dinner: A phyto-centric exploration of being edible', *Botanical Ontologies*, University of Oxford Conference. [online] Available at: https://www.academia.edu/7091917/Conversations_over_dinner_a_phyto_centric_exploration_of_being_edible. Accessed March 2022.

Attala, L., 2017. 'The edibility approach: using edibility to explore relationships, plant agency and the porosity of species' boundaries', *Advances in Anthropology* 7, 125–45.

Attala, L., 2019. '"I am apple." Relationships of the flesh. Exploring the corporeal entanglements of eating plants in the Amazon', in *Body Matters: Exploring the Materiality of the Body*, ed. L. Attala and L. Steel. Cardiff: University of Wales Press, pp. 39–61.

Baluska, F., Mancuso, S., Volkmann, D., and Barlow, P.W., 2009. 'The "root-brain" hypothesis of Charles and Francis Darwin: Revival after more than 125 years', *Plant signaling and behavior* 4(12), 1121–7, doi: 10.4161/psb.4.12.10574.

Baluska, F., and Mancuso, S., 2009. 'Plant neurobiology: from sensory biology, via plant communication, to social plant behavior', *Cognitive Process* 10 (Suppl 1), S3–S7.

Baldwin, I. T., Stork, W. F., and Weinhold, A., 2011. 'Trichomes as dangerous lollipops: Do lizards also use caterpillar body and frass odor to optimize their foraging?', *Plant signaling and behavior* 6(12), 1893–96.

Barad, K., 2003. 'Posthumanist performativity: Toward an understanding of how matter comes to matter', *Signs: Journal of Women in Culture and Society* 28(3), 801–31.

Barad, K., 2007. *Meeting the Universe Halfway: Quantum Physics and the Entanglement of Matter and Meaning*, Durham, NC: Duke University Press Books.

Barad, K., 2010. 'Quantum entanglements and hauntological relations of inheritance: dis/continuities, spacetime enfoldings, and justice-to-come', *Derrida Today* 3(2), 240–68.

Bateson, G., 1972. *Steps to an Ecology of the Mind: Collected Essays in Anthropology, Psychiatry, Evolution, and Epistemology*. Chicago, IL: University of Chicago Press.

Bateson, G., 1979, *Mind and Nature: A Necessary Unity*. London: Wildwood House.

Bateson, N., 2015. *An Ecology of Mind: Remember the Future – A Daughter's portrait of Gregory Bateson*. [online] Available at: *https://vimeo.com/ondemand/bateson?autoplay=1*. Accessed February 2022.

Bateson, N., 2018. 'Warm Data', *The International Bateson Institute*. [online] Available at *https://batesoninstitute.org/warm-data/*. Accessed February 2022.

Bennett, J., 2010. *Vibrant Matter: A Political Ecology of Things*. Durham, NC and London: Duke University Press.

Beyer, S. V., 2010. *Singing to the Plants: A guide to mestizo shamanism in the upper Amazon*. Albuquerque, NM: University of New Mexico Press.

Biedrzycki, M. L., and Bais, H. P., 2010. 'Kin recognition in plants: a mysterious behaviour unsolved', *Journal of Experimental Botany*, 61(15), 4123–8, doi: 10.1093/jxb/erq250.

Bird-David, N., 1990. 'The giving environment: Another perspective on the economic life of hunter gatherers', *Current Anthropology* 31(2), 189–96.

Bird-David, N., 1992. 'Beyond the hunting and gathering mode of subsistence: Observations on the Nayaka and other modern hunter-gatherers', *Man* 27, 19–44.

Blair, N., 2019. 'Lilyology as a transformative framework for decolonizing ethical spaces within the Academy', in *Decolonizing Research: Indigenous Storywork as Methodology*, ed. J. Archibald, J. Lee-Morgan and J. de Santolo. London: Bloomsbury Publishing, pp. 203–23.

Boserup, E., 1970. *Women's Role in Economic Development*. London: George Allen and Unwin.

Cahill, J. F., McNickel, G. G., Haag, J. J., Lamb, E. G., Samson, M., Nyanumba, M. and Cassady St Clair, C., 2010. 'Plants integrate information about nutrients and neighbors', *Science* 328, 1197–9, doi: 10.1126/science.1189736.

Coard, R., 2019. 'Done and dusted: Exploring the mutable boundaries of the body', in L. Attala and L. Steel (eds), *Body Matters: Exploring the Materiality of the Body*. Cardiff: University of Wales Press, pp. 157–72.

Crepy, M. A. and Casal, J. J., 2015. 'Photoreceptor-mediated kin recognition in plants', *New Phytologist* 205(1), 329–38, doi: 10.1111/nph.13040.

Devin, L., 2020. *The Forest Has You: An Initiation.* Castelvecchi: LIT Edizioni. [online] Available at: http://www.luisdevin.com/books/the-forest-has-you/. Accessed February 2022.

Diamond, J., 1987. 'The worst mistake in the history of human race', *Discover Magazine* (May), 64–6.

Douglas, M., 1966. *Purity and Danger: An Analysis of Concepts of Pollution and Taboo.* London: Routledge.

Dudley, S. A. and File, A. L., 2007. 'Kin recognition in an annual plant', *Biology Letters* 3(4), 435–8, doi: 10.1098/rsbl.2007.0232.

Gagliano, M. and Renton, M., 2013. 'Love thy neighbour: Facilitation through an alternative signalling modality in plants', *BioMed Central Ecology* 13(19), doi: 10.1186/1472-6785-13-19.

Gagliano, M., 2013a. 'Green symphonies: A call for studies on acoustic communication in plants', *Behavioural Ecology* 24(4), 789–96.

Gagliano, M., 2013b. 'The flowering of plant bioacoustics: How and why', *Behavioural Ecology* 24(4), 800–1.

Gagliano, M., 2015. 'In a green frame of mind: Perspectives on the behavioural ecology and cognitive nature of plants', *AoB Plants* 7, doi: 10.1093/aobpla/plu075.

Gagliano, M., 2018. *Thus Spoke the Plant: A Remarkable Journey of Groundbreaking Scientific Discoveries and Personal Encounters With Plants.* Berkeley, CA: North Atlantic Books.

Gagliano, M., Renton, M., Depczynski, M., and Mancuso, S., 2014. 'Experience teaches plants to learn faster and forget slower in environments where it matters', *Oecologia* 175, 63–72, doi: 10.1007/s00442-013-2873-7.

Gilbert, S. F., 2017. 'Holobiont By Birth: Multilineage Individuals As The Concretion Of Cooperative Processes', in *Arts of Living on a Damaged Planet*, eds A. Tsing, H. Swanson, E. Gan and N. Bubandt. Minnesota: University of Minnesota Press, pp. M73–M90.

Govier, E. and Steel, L., 2021. 'Beyond the "thingification" of worlds: Archaeology and the New Materialisms', *Journal of Material Culture* 26(3), 298–317, doi: 10.1177/13591835211025559.

Haraway, D., 2003. *The Companion Species Manifesto: Dogs, People and Significant Otherness.* Chicago, IL: Prickly Paradigm Press.

Haraway, D., 2016. *Staying with the Trouble: Making Kin in the Chthulucene*. Durham, NC and London: Duke University Press.

Hurn, S., 2012. *Humans and Other Animals: Human-Animal Interactions in Cross-Cultural Perspective*. London: Pluto Press.

Ingold, T., 2010a. *Bringing Things Back to Life: Creative Entanglements in a World of Materials*. NCRM Working Paper. Realities/Morgan Centre, University of Manchester. [online] Available at: https://hummedia.manchester.ac.uk/schools/soss/morgancentre/research/wps/15-2010-07-realities-bringing-things-to-life.pdf. Accessed February 2022.

Ingold, T., 2010b. 'The textility of making', *Cambridge Journal of Economics* 34(1), 91–102.

Ingold, T. (ed.), 2011. *Redrawing Anthropology: Materials, Movements, Lines*. Farnham: Ashgate Publishing.

Ingold, T., 2012. 'Toward an ecology of materials', *Annual Review of Anthropology* 41, 427–42.

Ingold, T., 2017. 'Taking taskscapes to task', in *Forms of Dwelling: 20 Years of Taskscapes in Archaeology*, ed. U. Rajala and P. Mills. Oxford and Philadelphia, PA: Oxbow Books, pp. 16–27.

Karban, R., Shoijiri, K., Huntzinger, M. and McCall, A. C., 2006. 'Damage-induced resistance in sagebrush: volatiles are key to intra- and interplant communication', *Oecologia*, 87(4), 922–30.

Karban, R., Shiojiri, K., Ishaizaki, S., Wetzel, W. C., and Evans, R. Y., 2013. 'Kin recognition affects plant communication and defence', *Proceedings of The Royal Society Biological Sciences*, 280:20123062.

Kimmerer, R., 2014. 'Returning the gift', *Minding Nature* 7(2), 18–24.

Kohn, E., 2007. 'How dogs dream: Amazonian natures and the politics of transspecies engagement', *American Ethnologist* 34(1), 3–24.

Kohn, E., 2013. *How Forests Think: Towards an Anthropology of Beyond the Human*. Berkeley, Los Angeles, CA and London: University of California Press.

Kohn, E., 2015. 'Anthropology of ontologies', *The Annual Review of Anthropology* 44, 311–27.

Latour, B., 2004. 'Whose cosmos, which cosmopolitics: comments on the peace terms of Ulrich Beck Symposium: talking peace with gods, part 1,' *Common Knowledge* 10(3), 450–62.

Luna, L. E., 1984. 'The concept of plants as teacher among four Mestizo shamans of Iquitos, Northeast Peru', *Journal of Ethnopharmacology* 11(2), 135–56.

Macfarlane, R., 2020. *Underland: A Deep Time Journey*. London: Penguin Random House.

Mancuso, S., 2021. *The Nation of Plants: A Radical Manifesto for Humans*. London: Profile Books.

Marder, M., 2014. *The Philosopher's Plant: An Intellectual Herbarium*. New York, NY: Columbia University Press.

Meents, A. K. and Mithöfer, A., 2020. 'Plant–plant communication: Is there a role for volatile damage-associated molecular patterns?', *Frontiers in Plant Science*, 11:583275, doi: 10.3389/fpls.2020.583275.

Milton, K., 2000. 'Hunter-gatherer diets – a different perspective', *The American Journal of Clinical Nutrition* 71(3), 665–7, doi: 10.1093/ajcn/71.3.665.

Morris, J. L., Puttick, M. N., Clark, J. W., Edwards, D., Kenrick, P., Pressel, S., Wellman, C. H., Yang, Z., Schneider, H., and Donoghue, P. C. J., 2018. 'The timescale of early land plant evolution earth', *Atmospheric, and Planetary Sciences* 115(10), E2274–E2283, doi: 10.1073/pnas.1719588115.

Morton, T., 2010. *The Ecological Thought*. Cambridge, MA and London: Harvard University Press.

Morton, T., 2013. *Hyperobjects: Philosophy and Ecology After the End of the World*. Minneapolis, MN and London: University of Minnesota Press.

Morton, T., 2016. *Dark Ecology: For a Logic of Future Coexistence*. New York NY: Columbia University Press.

Mosko, M. S., 1987. 'The symbols of "forest": A structural analysis of Mbuti Culture and social organisation', *American Anthropologist* 89(4), 896–913.

Narby, J., 1999. *Cosmic Serpent: DNA and the Origins of Knowledge*. New York, NY: Tarcher/ Putham Paperbacks.

Narby, J. and Chanchari Pizuri, R. C., 2021. *Plant Teachers: Ayahuasca, Tobacco and the Pursuit of Knowledge*. Novato, CA: New World Library.

Nast, H. J., 2006. 'Loving … whatever: Alienation, neoliberalism and pet-love in the twenty-first century', *ACME: An International E-Journal for Critical Geographies* 5(2), 300–27.

Pollan, M., 2002. *The Botany of Desire: A Plant's Eye View of the World*. London, Berlin and New York, NY: Bloomsbury.

Pollan, M. 2013. 'The intelligent plant', *The New Yorker*. [online] Available at: *http://www.newyorker.com/magazine/2013/12/23/ the-intelligent-plant*. Accessed February 2022.

Ravalec, V., Mallendi, and Paichler, A., 2007. *Iboga: The Visionary Root of African Shamanism*. Paris: Park Street Press.

Reichel Dolmatoff, G., 1985a. *Los Kogi* I. Bogotá: Procultura S.A.

Reichel Dolmatoff, G., 1985b. *Los Kogi* II. Bogotá: Procultura S.A.

Rival, L., 1993. 'The growth of family trees: Understanding Huaorani perceptions of the forest', *Man* 28(4), 635–52.

Sagan, D. and Margulis, L., 1988. *The Garden of Microbial Delights: A Practical Guide to the Subvisible World*. Boston, MA, San Diego, CA, New York, NY: Harcourt Brace Jovanovich.

Sahlins, M. D., 1974. *Stone Age Economics*. Chicago, IL and New York, NY: Aldine Atherton Inc.

Sapir, E., 1929. 'The Status of Linguistics as a Science', *Language* 5, 207–19.

Schaefer, H. M. and Ruxton, G. D., 2011. *Plant-Animal Communication*. Oxford: Oxford University Press.

Schwartz, B., 2015. *Why we Work*. TEDBooks London: Simon and Schuster.

Schwartzberg, L., 2014. 'Hidden miracles of the natural world', *TED talks*. [online] Available at: *https://www.ted.com/talks/louie_ schwartzberg_hidden_miracles_of_the_natural_world*. Accessed February 2022.

Simard, S. W., 2009a. 'Mycorrhizal networks and complex systems: Contributions of soil ecology science to managing climate change effects in forested ecosystem', *Canadian Journal of Soil Science* 89(4), 369–82.

Simard, S. W., 2009b. 'The foundational role of mycorrhizal networks in self-organisation of interior Douglas-Fir forests', *Forest Ecology and Management* 258S, S95–S107.

Simard, S. W., 2012. 'Mycorrhizal networks and seedling establishment in Douglas-fir forests', in *Biocomplexity of Plant–Fungal Interactions*, ed. D. Southworth. Oxford: John Wiley & Sons, pp. 85–107.

Simard, S. W., 2021. *Finding the Mother Tree: Uncovering the Wisdom and Intelligence of the Forest*. London: Allen Lane.

Simard S. W., Philip, L. J. and Jones, M. D., 2011. 'Pathways for belowground carbon transfer between paper birch and Douglas-fir seedlings', *Plant Ecology and Diversity* 3(3), 221–33.

Simard, S. W., Beiler, K. J., Bingham, M. A., Deslippe, J. R., Philip, L. J. and Teste, F. P., 2012. 'Mycorrhizal networks: mechanisms, ecology and modelling', *Fungal Biology Reviews* 26(1), 39–60.

Steffensen, V., 2020. *Fire Country: How Indigenous Fire Management Could Help Save Australia*. Melbourne and Sydney: Hardie Grant Publishing.

Stern, L., 2017. 'A garden or a grave? The canyonic landscape of the Tijuana-San Diego region', in *Arts of Living on a Damaged Planet*, eds A. Tsing, H. Swanson, E. Gan and N. Bubandt. Minneapolis, MN: University of Minnesota Press, pp. G17–G29.

Strathern, M., 1988. *The Gender of the Gift: Problems with Women and Problems with Society in Melanesia*. Berkeley, CA: University of California Press.

Taiz, L., Alkon, D., Draguhn, A., Murphy, A., Blatt, M., Hawes, C., Theil, G. and Robinson, D. G., 2019. 'Plants neither possess nor require consciousness: Opinion', *Trends in Plant Science* 24(8), 677–87.

Taussig, M., 1992. *The Nervous System*. New York NY and London: Routledge.

Thambinathan, V. and Kinsella, E. A., 2021. 'Decolonizing methodologies in qualitative research: Creating spaces for transformative praxis', *International Journal of Qualitative Methods*, 20, 1–9.

Tuan, Y., 1984. *Dominance and Affection: The Making of Pets*. New Haven, CN and London: Yale University Press.

Tuhiwai Smith, L., 2021. *Decolonizing Methodologies: Research and Indigenous Peoples*. London: Bloomsbury Publishing.

Turnbull, C. M., 1994. *The Forest People*. New York, NY: Touchstone.

UNESCO-BRIDGES, 2019. *Toward the establishment of Bridges: action to promote sustainability science*. [online] Available at: https://en.unesco.org/news/toward-establishment-bridges-action-promote-sustainability-science. Accessed 28 March 2022.

Note

1. Also see the work of Gagliano et al. (2014) who demonstrated that *Mimosa pudica* remembered and therefore could learn.

2 THE MATERIALITY OF PLANTS
Plant–People Entanglements[1]
Marijke Van der Veen

Introduction

Materiality is a concept now widely used in archaeology to explore relationships between people and objects, encouraging a move away from a focus on what objects 'mean' to what they 'do' and what they 'are'. The exact use and definition of 'materiality' has changed over time, from an initial focus on the agency of objects, that is the recognition of 'material and nonhuman agency' (Knappett and Malafouris 2008, ix–xix; see also, e.g., Gosden 2005), to a greater emphasis on relationships, networks and meshworks, rather than on the individual agency of either person or object (e.g., Ingold 2007; Knappett 2011; Robb 2010). Increasingly, animals are brought into these discussions too, with a desire to move away from anthropocentric approaches within zooarchaeology (e.g., Overton and Hamilakis 2013), but plants have yet to feature prominently, though see Fuller, Allaby and Stevens (2010); Head, Atchison and Gates (2012), and Jones and Cloke (2008). Here I seek to explore the materiality of plants and to highlight their centrality in all aspects of human interactions. I am taking a deliberately broad view of what 'materiality' means, considering both the material properties of plants *and* the material and social implications of plant–people relations. After a quick look at 'what plants want', the various aspects of plant–people interconnections are discussed under the headings of embodiment, investment, praxis and entanglement.

What do plants want?

Plants are essential to human and animal life on earth: they create the oxygen we breathe and the food we consume. Additionally, plants provide the fibres for our clothes, the building materials for our shelter, the fuel that keeps us warm, and the flowers that give us beauty.

Following previous studies by Mitchell (1996) and Gosden (2005), it is helpful to consider briefly: 'what do plants want?' The answer is quite simple. Fundamentally, the primary desire of plants is to reproduce, to make copies of themselves, metaphorically speaking.[2] A variety of strategies has evolved to achieve this. For example, some (many trees) rely on the wind to disperse their pollen and produce large quantities of pollen to increase the chance of successful pollination. Others (many flowers) use brightly coloured petals or a sweet smell and ample nectar to attract bees and other insects to visit them and aid pollen dispersal, thus requiring less pollen. In contrast, many cereals do not rely on others for fertilisation, but self-pollinate, while yet other plants (including weeds) can self-propagate from root fragments. Some produce seeds or nuts that are edible, ensuring that small animals are happy to eat and disperse them, others have developed seeds with aerodynamic properties to allow the wind to carry them, or burs that attach themselves to the coats of passing animals. Plants also use chemical warfare; they use phytochemicals (so-called secondary compounds) to ward off predators (poisonous substances, unpleasant smell or flavour), or to attract pollinators (e.g., pigments). Some of the phytochemicals and other material aspects of plants have become very attractive to humans (poison, medicine, mind-altering substances, as well as food, fibres, sweet smell and beauty). Furthermore, the fact that some of the properties that originally evolved to deter herbivores have subsequently been embraced by some humans as pleasurable or desirable (e.g., herbs, spices), highlights that these properties exist only within the context of specific relationships and entanglements, see below. The most fundamental reason why plants produce these chemicals and strategies is because they cannot locomote. They cannot run away from predators or move to a nearby plant to achieve fertilisation, nor can they move outside of their region of origin. To overcome this disadvantage, they have developed strategies to get others to help them to do so, and these agents include wind, insects, birds, and other animals, as well as humans.

Various authors have emphasised how the historic relationship between people and plants is often reciprocal, one of co-evolution. For example, Pollan (2001) highlights how the domestication and selective breeding of plants – the examples he uses are the apple, the tulip, marijuana and the potato – did not only benefit people, but plants too, in that their primary desire, to make copies of themselves, was met, in that it facilitated and enhanced their reproduction and proliferation.

> Through trial and error these plant species have found that the best way to do that [reproduce themselves] is to induce animals – bees or people, it hardly matters – to spread their genes.
> (Pollan 2001, xiii)

By playing on and enabling our desire for sweetness, beauty, intoxication or control, conscious or otherwise, the plants have been able to reproduce and multiply and spread across the globe. Thus, while the process of domestication has often been regarded as a process brought about by people, we might equally consider it something that plants and animals have done to us, 'a clever evolutionary strategy for advancing their own interests' (Pollan 2001, xiv). After all, 'domesticated' species such as wheat, potatoes, apples, tulips, sheep and dogs are much more common today, present over much larger geographical areas and in much greater variety than their original wild ancestors (Pollan 2001, xv).

Crops are not the only group of plants that have entered into a mutualistic relationship with people; weeds are another. In ecology they are referred to as synanthropes: wild species that live near people and benefit from a close relationship with people and their artificially created habitats – equivalents in the animal world are grain beetles, head lice and rats. They tend to be perceived as pests; for example, despite the large-scale use of herbicides in modern agriculture, weeds still reduce arable productivity by 10 to 20 per cent (Mabey 2010, 4). There are many definitions of weeds; most describe weeds from the point of view of farmers and gardeners, that is, as plants growing where they are not wanted, but a broader ecological view sees them as: plants that contest with man for the possession of the soil (Blatchley 1912), as opportunistic species that follow human disturbance of the habitat (Pritchard 1960), as artefacts or camp followers (Anderson 1954), as pioneers (Bunting 1960) or as plants uniquely able to thrive in land subject to the plough (Isely 1960; for references, see Harlan 1992, Table 4.1). In all cases the link between weeds and people is clear: 'the history of weeds is the history of man' (Anderson 1954, as quoted in Harlan 1992, 84); 'the decision to cultivate also meant having to weed' (Pollan 1989).

Weeds are important in archaeology, because of their close association with arable crops. Their life form (annual/perennial) and ecology (preference for nutrient rich/poor, acid/neutral and wet/dry soils) help to identify the conditions in the arable fields, from which cultivation techniques, arable strategies (manuring, weeding, ploughing, etc.) and scales of production can be inferred. Significantly, these synanthropes also highlight how problematic the nature/culture dichotomy, so prevalent in our western thinking, actually is. Weeds are wild species, but they live in cultural, i.e., human-created, habitats. Their history is bound up with their mutual relationships with people and human practices, just as much as that of 'domesticates' is. Thus, to regard 'crops' as pertaining to the world of 'culture' and 'weeds' as unwanted intruders from an external world of 'nature', misses the point entirely – people, crops and weeds (as well as animals, artefacts, etc.) all occupy the same, mutually constructed ontological realm.

Weeds also apply various strategies for survival and reproduction. Annuals with short growth cycles produce many seeds; even if they are weeded out during the growing season of the cereal crop, they have usually already cast some seed – or seed from a previous year still lies dormant in the soil – and they can thus quickly reproduce. Some perennial weeds, such as bindweed, have underground rhizomes from which they can regenerate and they have become pernicious weeds in fields and gardens. Pollan (1989) describes it as follows: 'the evolution of bindweed took the hoe into account. By attacking it at the root, we play right into its insidious strategy for world domination.' Others are so-called 'professional' weeds, such as wild oat, brome grass, darnel and false flax; they mimic the crop in terms of growth cycle, seed size and shape, so that their seeds are harvested with that of the crop, and sown with it again at the start of the next season, thus ensuring reproduction (Harlan 1992, 93).

People have developed needs and desires for food, flavour, beauty, intoxication, fibres, resins, dyes, timber and shade – all properties and materials that plants can provide. Plants use those human (and non-human animal) needs to achieve their own desire, that is, to reproduce themselves. Those that have made themselves the most attractive, have been the most successful in reproducing and spreading themselves. While the relationship was often one of mutual benefit, it also altered both the plants and the people. Biologists tend to use the term 'coevolution' to describe this process; in archaeology the terms entanglement and relationality are more common, as discussed below.

Embodiment – you are what you eat

All food we eat is either directly or indirectly derived from plants; thus we consume plants on a daily basis. In contrast to the consumption of objects, the consumption of plants also involves the physical ingestion into the body; they become incorporated and embedded into the body. This material aspect of the body is now a growing research area within archaeology (e.g., Robb and Harris 2013; Sofaer 2006).

The first, and at face value the most immediate, impact of plants on people may be that of food and their nutritional value. The ingestion of plant foods has a physical effect on us, it maintains life, just as the lack of food, or the lack of healthy foods, can make us ill or lethargic. The mild sedative properties of some plants help us to relax or sleep, while others help us to stay awake and alert, or have other properties in terms of assisting in the healing of wounds, knitting bones together, clearing the airways, tackling infections, etc. Having access to enough and adequate quality food, and being able to share food, form significant components of our wellbeing, physically, mentally and socially.

The material effect of food ingestion can be studied through the chemical signature that certain classes of food leave behind in the body. Most of the 'you are what you eat' research in archaeology focuses on two broad aspects: what was eaten (together with the associated aspects of subsistence, agriculture and trade) and the social/cultural role of foods. For example, stable isotope analysis uses chemical signatures left in the bones by the consumption of certain foods (plant and animal based) to determine diet, while archaeobotany and zooarchaeology use the actual food preparation and consumption left-overs to do the same. Both also study social and cultural aspects of food, such as differential access to certain foods (especially meat and exotics), and the use of food as an indicator of, or a means to acquire, a specific identity (social, ethnic, religious, etc.). Both aspects are immensely valuable, offering information on changes in subsistence, agricultural strategies, long-distance trade, different consumer groups, diversity within the population, etc. But the emphasis is firmly on how people used the plants (and animals), that is, on plants as objects.

A second type of impact concerns plants and plant substances that have addictive or even mind-altering properties. The consumption of certain stimulants and intoxicants, such as alcohol, tobacco, cannabis and opium, but also substances such as sugar, and to a lesser extent tea and coffee, can result in forms of addiction; their consumption has us (or at least some of us) craving for more. These plant substances all tend to affect our chemical make-up; they alter our physiological and our psychological being. This may result in a relationship of dependency, which can and does have significant material effects on the lives of those afflicted, both in terms of their physical and mental well-being and thus their social life. Yet, how do we distinguish between the various types of plant substances? In western society we differentiate between plants that are nutritious, remedial, psychoactive and toxic, but these categories are not always clear-cut. A small amount or low concentration may be remedial or give pleasure, while a large amount or regular intake can be toxic or addictive. Furthermore, 'nutritious' foods can also affect one's mental state; the concept of a separate mind and body is problematic.

Studies dealing with these effects of ingestion are still rare in archaeology, despite the fact that psychoactive substances are an integral part of human culture and social life (Goodman, Lovejoy, and Sherratt 1995). Most of the work on alcohol, for example, focuses on its role in feasting, in ritual and as a social lubricant, and in-depth studies on mind or behaviour altering substances and the contexts in which these were used by people, are few; though see Lewis-Williams (2004). These issues may seem unrelated to the relationships usually considered under the umbrella of 'materiality', but, as we shall see below, the human desire for certain foods and plant substances and the plants' desire for reproduction have created complex entanglements between plants and people.

Investment – seeds as archives

A reverse form of embodiment is one where human/plant interactions have left a signature in the physical make-up of the plant ('emplantment' instead of 'embodiment'). Two examples spring to mind, one relating to the impact of agricultural practices, the other to the effect of plant selection and mutation in the process of crop dispersal. For example, it is now evident that certain types of soil improvements, i.e., the manuring and irrigation of arable fields, leave a chemical signature in the seeds of the crops grown in those fields; that is, manuring results in high nitrogen values ($\partial 15N$), while water availability and irrigation can be inferred from stable carbon values ($\Delta 13C$) (Fraser et al. 2013; Wallace et al. 2013). Here the relationship between people and plants is one of mutual benefit. When the growing conditions in the fields are improved, the crops benefit by being able to survive, thrive and reproduce in environments otherwise marginal or hostile to them. Other plants, i.e., weeds, will take advantage too, grateful for the new availability of suitable habitats. People are rewarded for their efforts through gaining higher or more reliable yields, while we archaeologists benefit too, in that both the chemical signatures in the actual crop seeds, and the types of weeds thriving in those fields (and harvested with the crops) provide us with valuable information about past people's relationships with plants, e.g., how they cultivated their crops, how much they invested in soil improvement, etc.

Similarly, the DNA of historical landraces of crops (and of archaeological seeds once we can successfully extract it) provides information about the spread of these crops out of their area of origin, and about the complex interconnected roles of plants, people and climate. For example, DNA analysis now points to multiple domestication events of several Near Eastern crops (e.g., barley, emmer and einkorn), which has implications for our understanding of the intentionality or otherwise of the domestication process (see below under 'Entanglement'). In the case of barley, this same evidence also suggests that this crop was introduced into Europe more than once. Current evidence identifies three separate groups of barley in Europe, each originating from a different part of South-West Asia, and showing a clear latitudinal distribution pattern (Jones et al. 2012, 2013). One of them is a so-called 'day-length non-responsive type', i.e., a strain of barley that can cope with a long growing season and wet summers, and this is the group found predominantly in north-west Europe. Archaeobotanical and genetic evidence combined suggest that this strain of barley was a later introduction into Europe, possibly in the fourth millennium BCE – simultaneous with other introductions and agricultural changes – having previously been domesticated in Iran. This offers an important new perspective in debates concerning the spread of agriculture into Europe and may help to explain why the spread of agriculture northwards into Europe was not continuous, slowing down at certain locations (north of the Hungarian plain, in the Alpine region, and in Scandinavia). Previous explanations for this phenomenon have included the relative success of the hunter-gatherer lifestyle, cultural preferences for certain foodstuffs, and the time needed for crops to adapt to new climatic conditions. This new DNA evidence suggests that the latter explanation, i.e., the introduction of a day-length non-responsive barley, is likely to have been one of the key factors involved, without, of course, ruling out additional factors (Jones et al. 2012, 2013). Thus, here too the material properties of the plants allow us to unravel the complex interactions between plants, climate and people. These properties of plants, as we have seen, are not fixed. They change in the context of their relation with people – and this process is mutual; people are changed too – and the properties of plants thus form archives of past human and plant behaviour.

Praxis – a way of life

Yet another form of embodiment is that generated by the routine, day-to-day engagement with plants, in the sense of gathering, tending, cultivating or growing plants (Ingold 1993). In the case of cereal farmers, their daily, monthly and yearly rhythms are tied to the life cycle of the cereals they grow. The soil needs to be prepared at a certain time of the year, just as do the acts of sowing, weeding, harvesting, threshing, and storing, and the tools of the trade and the movements made with those tools (spade, plough, traction animals, scythe, threshing stick or sledge, sieves) are fixed too. This engagement with the process of farming, enacting the same set of actions over and over again, year after year, makes farmers who they are. By doing it they become farmers, a particular mode of being, an ontology that is rooted not in some unchanging essence but in particular, historically arisen relationships that are in a continuous process of becoming.[3] These embodied routines condition how farmers see and interact with the world, the landscape, the plants and the animals, as well as other humans; they are their life. A good illustration of this identification of the work with the life is given by Harlan (1992, 8). A farmer and his family were seen harvesting a field in eastern Turkey. Asked why he did not use a scythe and cradle, as that would halve the time needed to harvest the field, the farmer answered: 'Then what would I do?' A farmer's engagement with the plants and the associated rhythms and communications with the world around him/her (plants, animals, tools, buildings and humans) is thus fundamentally different from that of, for example, the hunter–gatherer. Both need to plan, to ensure enough food for family and future needs, but the necessary time management and scheduling varies considerably, as does their engagement with material culture (Bradley 2004).

In contrast, gardening and horticulture have different rhythms. Here the focus tends to be on root and tuber crops, vegetables, herbs and fruits – cereals can also be grown in small garden plots – grown mixed together in the same garden, in contrast to large fields with one or two annual crops. Many of these plants thrive under intensive cultivation with regular monitoring and nurturing, using a strategy of 'little and often', and garden plots are thus often located close to the home. The plants are, consequently, nearer and the plant-people relations closer, incorporating more handling of individual plants, such as planting, division and taking cuttings, as well as weeding, watering and feeding (Hastorf 1998; Leach 1997). This closer familiarity between cultivator and plant may have given rise to greater experimentation and the development of new varieties, and may have generated a more prominent role for plants in people's ritual life, medicine, etc. The proximity of the garden plot to the house may help to explain why gardening is perceived by some as a 'domestic' activity, carried out foremost by women. Here the intimacy of the daily tending of the plants evokes similarities with that of rearing or nurturing children. There is, in fact, considerable ethnographic evidence for gender differences between farming systems, with women more actively associated with hoe-based cultivation (gardening, horticulture), and men with plough-based agriculture (field crops). This foregrounds dissimilarity between nurturing or mothering and husbandry, levels of engagement and associated rhythms (e.g., Hastorf 1998; Ingold 1996; Leach 1997; Sherratt 1981).

Each type of crop will bring its own rhythm. For example, several so-called summer crops were introduced into the Middle East, North Africa and Spain during the Islamic period (AD 700–1200), including citrus fruits, sugar cane, cotton, rice, and banana (Van der Veen 2011; Watson 1983). These are tropical or sub-tropical crops that originate in South or South-East Asia and/or the Pacific and require high temperatures, plenty of sunlight and large amounts of water. The lack of sufficient rainfall during the summer months in North Africa and the Middle East meant that irrigation was an essential component of their cultivation here. Not only did this require a considerable investment in terms of infrastructure (digging and maintaining irrigation canals, building and maintaining water wheels, arranging access to the available water; fertiliser), it also drastically changed the agricultural year and, thus, the lives of the farmers. Previously, the emphasis had been on temperate crops, with the key staples (cereals, pulses and oil/fibre crops) grown during the winter and harvested in spring. The summer had been the quieter season of the agricultural calendar, with much of the land lying fallow, except for some cultivation of fruit and vegetables, using irrigation on a small scale. By the tenth or eleventh century large-scale irrigation had become commonplace. For example, al-Masudi (tenth century) refers to the Egyptians as the richest of all men in sugar and al-Idrisi (twelfth century) describes the area around Cairo as one well-watered field of sugar cane (Watson 1983, 28). No longer was the summer a period of slowing down; the agricultural year now lasted all year round, greatly increasing farmers' workloads and practices. Large-scale irrigation is often paired with largescale land ownership, and several of the new crops lend themselves to a form of plantation or 'industrial' cultivation. Thus, lives will have changed too, both in terms of the new annual rhythms, different tools and techniques used and embodied, but also in terms of new labour relations (Van der Veen 2011, 231–3).

Fruit tree cultivation brings its own conditions. Trees are long-lived plants – lasting as long as or longer than a human generation – and they can take up to ten years to start fruiting. The produce then arrives all at the same time of the year, while often not being suitable for long-term storage, unlike the cereals – gardeners with apple or plum trees will recognise the annual glut of fruit. Here, farmers engage not just with the demands for long-term planning, investment and different practices needed when growing trees (time, land, finance, know-how, pruning, grafting), but also with the need for an outlet for their produce, which is why orchards and market gardens are often found in close proximity to centres of population such as towns. Thus, the reality of being a cereal farmer, a gardener, a landowner growing sugar cane or a horticulturalist growing fruit for a market, is directly tied into the materiality of the crops they grow and look after. Their lives are regulated by the life cycles, the needs and the idiosyncrasies of their plants. A simple dichotomy between hunter-gatherer and farming lifestyles is thus unhelpful.

Entanglement – webs of relationality

So far, the plant-people relations have been described as mutualistic, as examples of co-evolution or co-dependency (terms used commonly by biologists), or as impacting on or creating new ways of living. The relationships often go further, however, and here the concepts of 'meshwork' and 'entanglement' are appropriate. Ingold (2007) coined the term meshwork to describe the interwoven lines along which materials flow, mix and mutate, in contrast to the term 'network' which implies connections between specific points. Hodder (2011, 164) uses the term 'entanglement' to try and capture the fact that both plants and humans become caught up or trapped in the relationships: 'humans get caught up in a double-bind, depending on things [plants] that depend on humans'.

An example of such entanglement is that of the transition to farming. For example, Harlan (1992, 3) regards cereal crops as 'artefacts', while at the same time recognising that people have become so dependent on these crop plants that, in some way, the plants have 'domesticated' people. 'A fully domesticated plant cannot survive without the aid of man, but only a minute fraction of the human population could survive without cultivated plants. Crops and man are mutually dependent' (Harlan 1992, 3). Harlan refers here to the fact that, in the process of domestication, the cereals and pulses lost their natural dispersal mechanisms (loss of the brittle rachis or pod dehiscence at maturity), making the plants dependent on human dispersal for their reproduction. The loss of natural seed dispersal in cereals is in many instances based on just a single mutation, and the selection for those mutants may have been the unintentional result of, for example, a change in the way the wild cereals were harvested. People may thus unwittingly have become entangled into a closer relationship with the plants, hence becoming trapped in a set of complex labour-demanding activities (Fuller, Allaby and Stevens 2010).

The 'materiality' of the plants (not a term Harlan used, of course) resulted in the farmers becoming bound to them, in that crops need to be rooted in one place till maturity, tend to produce fruit at a fixed time of the year, and need care and attention before and during the growing season. While sedentism is not a pre-requisite for agriculture – small plots of land can be and are cultivated by primarily mobile groups – it often leads to or flows from agriculture. One of the more significant aspects of the transition to farming is the fact that cereal grain (and certain other crops, e.g., pulses) can be stored for long periods, allowing permanent settlement and the creation of a food surplus, which could be used to alleviate periods of shortage, but also facilitated periodic feasting and gift exchange. This, in turn, expedited a new and more abundant and permanent material culture, wealth accumulation and the rise of social inequality, with ownership of land becoming a key factor in social relations within larger scale farming systems, and the emergence of different and asymmetrical modes of being, e.g., those of the landlord and the tenant farmer. The payback would have been increased reliability of harvest and yield and increased control over resources, while it allowed the plant to spread and reproduce beyond its natural habitat. But the changes in vegetation, fauna, landscape, material culture and social relations meant that often there was no way back to the old way of life (e.g., Robb 2013). The transition to farming and the associated emergence of sedentism, ownership and wealth accumulation, may be regarded one of the best examples of the people-plant entanglement, in that it brought about fundamental changes in society, vegetation and the material world.

Such entanglements can be seen in many other situations, for example when new crops are introduced to new regions. The introduction of summer crops into the Middle East, referred to above, is a case in point. Watson (1983) identified complex interdependencies (entanglements or meshworks we would now say) between the introduction of the new crops, new agricultural techniques (irrigation, water wheel, greater use of crop rotation, fertiliser), changes in land tenure, the unification of the Middle East under Islam (political stability, greater mobility), economic growth, rapid urbanisation, the flourishing of Islamic culture, the unusual receptivity of early Muslim civilisation to all that was new, and an eagerness by the new wealthy classes to learn and improve their lives. The new crops offered rich rewards: many were potential cash crops (sugar cane, cotton, banana, citrus fruits), and those living close to the new urban centres could profit from selling their produce to these emerging and growing markets. Yet, it also enhanced exposure to taxation and created a greater dependency on those markets, just as too great a distance from the new markets or from water for irrigation meant not being able to participate in the new prosperity. The specific properties of the plants, combined with local environment and socioeconomic developments created opportunities, as well as dependencies, inequalities and greater regional diversities (Van der Veen 2011, 231–3).

The history of sugar shows that its impact and entanglement did not stop with its introduction into the Middle East (see Mintz 1985). From there the consumption of sugar quickly spread to Europe. Initially it was an expensive and exotic luxury, but the demand for cheap sugar led to its introduction into the Caribbean and the creation of large plantations there. It is a very labour-demanding crop and the price could only be driven down with the help of cheap labour, in the form of slaves. The local population, not resistant to the new diseases brought from Europe, was quickly decimated and over time they were replaced by slaves from Africa, as part of the transatlantic slave trade. The African slaves made sugar cheap to produce and thus profitable, and they also became the dominant ethnic group in the Caribbean, while the new plantations caused the destruction of the region's rainforests (Viola 1991). This mode of production meant that sugar became cheap enough to be combined with tea, offering factory owners in Britain the ideal break-time 'food' (energy plus stimulant) to sustain their workforce during their long working hours (Mintz 1985, 149), and even today its cheapness and addictive properties impact on our lives (obesity). Further entanglements may be clear from the fact that the introduction of maize from America into West Africa, while providing a reliable and nutritious staple there, thus sustaining population growth, may have provided additional African manpower for the American plantations that produced sugar, cotton and tobacco (Viola 1991). Other examples of plants that have had similarly long-lasting effects on human society, linked as they are to slavery and/or colonialism are tea, cotton, coffee, cocoa and potato, but also the medicine quinine which, by alleviating the effects of malaria, allowed Europeans to settle in parts of Africa and Asia (Hobhouse 1992).

A very different effect of newly introduced food plants concerned their ability to raise strong emotional reactions, and the archaeology of emotions is a growing field of study (e.g., Gosden 2004; Harris and Sørensen 2010; Tarlow 2000). The close links between food, emotions and social relations mean that plant foods play a powerful role in our social worlds. Just as groups of related objects may shape new social entities and relationships (e.g., ceramics: Gosden 2005), so can new foods. These may be associated with a foreign culture, for example the Mediterranean foods brought into north-west Europe during the Roman conquest of this region. In Roman Britain this concerned some fifty food introductions, ranging from exotics such as black pepper and Mediterranean foods such as grapes, olives, various herbs and wine, to what are now common 'English' fruits and vegetables, such as apples, plums and cabbage (Van der Veen 2008). Their physical presence and otherness will have demanded an emotional reaction: one of curiosity and delight in the new tastes, one of indifference, one of disgust and aversion, or one of economic and social opportunity. In the first instance the new foods will have been associated with the new ruling power and this will have had a social effect too, with some individuals and groups of individuals desiring access to the new foods in order to belong to the new elite and/or feeling the need to change their foodways in order to belong, and with others feeling a sense of exclusion, inferiority or inadequacy for not being able to afford the new and expensive foods or not knowing how to eat them. Others may have felt antipathy, wanting to spurn things foreign or rejecting everything associated with the alien culture. While these new foods were initially part of a foreign culture, they did not necessarily stay that for long. Some were quickly incorporated into local agriculture and foodways (Van der Veen 2008). The introduction and physical presence of these new foods created the need for new cultivation practices and tools, and forced people to make choices and new alliances, thus taking part in the changing social and material realities of life in Roman Britain.

The 'materiality' of plants seems very obvious in these examples, affecting both production and consumption processes as well as social relations on a vast scale. Plants affect the physical and psychological well-being of those who consume them, and they register the selections and treatments received. Their ability not only to entice people but also to mutate – and thus to thrive successfully in new environments

– combined with people's desire for their nutrition, beauty or many other properties (stimulants, fibre, etc.) has created complex entanglements and varied webs of relationships. Plants have come out of this entanglement well; they may have lost their freedom, their reproduction mechanisms (at least in the case of certain crops), but they have gained an increased reproduction rate and a spread over much larger parts of the globe than their wild predecessors. People have gained too, through access to a greater variety of food and flavours and plant fibres, more control over food, a wider range of nutrients, and access to a wider range of animal products, through growing fodder crops. But their lives were altered too: less freedom, sedentism, ownership, living in larger groups, altered social structures, becoming enslaved to certain foods, or tied into exploitative modes of production. While humans may, up to a point, have become willing slaves in the plants' reproductive processes, some crops literally enslaved people.

Discussion and conclusion

Plants are central to human life and people-plant relationships have been central to archaeology and environmental archaeology from the start. Apart from questions such as 'what did they eat?' and 'what was the vegetation like?', research in this area has focused on three broad aspects: how people were able to gain control over nature (domestication, field systems, secondary products, soil improvement, colonisation of marginal land, etc.), how they damaged it (deforestation, soil erosion, loss of bio-diversity), as well as how they utilised plants (and animals) in the creation of identity and social relations (feasting, social status, food avoidance). Plants were studied from an anthropocentric point of view, understandably so in the context of archaeology rather than biology. What I have tried to do in this paper, however, is to bring into clearer focus the fact that plants have agency too; that they affect us, as much as we affect them. All gardeners will know that plants have a will of their own. We only have to turn our back for a few weeks and plants are growing where we did not want them to, new ones have arrived, and others have strangled or swamped their neighbours. Plants have their own agenda, that is, to make copies of themselves, and will use a variety of strategies to achieve this. As Pollan (2001) so compellingly illustrates, plants often have something we desire, be that nutrition, sweetness, beauty, or intoxication, but in multiplying their produce we not only directly assist the plants in their desire to reproduce, but we also become affected by them in ways that we never could have envisaged, including the emergence of these very desires. These entanglements between plants and people are felt in the arenas of consumption and production, social life, labour relations, intercontinental contacts, body and mind. Plants are part of our everyday lives; they assist us, define us and create historical associations. Farmers are partly the crops they grow, tomatoes (of New World origin) are intricately linked with Italy, tulips (of SW Asian origin) with Holland, potatoes (of New World origin) in part define the history of Ireland, poppies remind us of Flanders Fields and the horrors of World War I, the apple symbolises the fall from grace in the Garden of Eden, rubber facilitates motorised transport and assists in the fight against AIDS, while sugar enslaved millions of people and has now become the curse of our modern world.

I have deliberately focused on plants in this paper, and used a few thumbnail examples, rather than detailed case studies, to illustrate the breadth of contexts in which plants impact on our lives. As a result, the entanglements with the environment, and with objects, buildings and material culture more widely, have only been briefly mentioned, and in future papers these need further elaboration. I am thinking here of the intricate interrelationships between plants and ecosystems (substratum, rainfall, temperature, living organisms), farm animals (ploughing, manure, transport, fodder), tools and technology (ploughs, mills, grindstones, water-lifting devices, olive/wine presses, carts), storage facilities (barns, granaries, ceramic vessels), processing areas (threshing floors, corn-driers, malting ovens), and trade (shipping, navigation, harbours) and consumption (hearth, table, communal feasting area, coffee shop, restaurant, opium den, pottery, cutlery), to name but a few. These meshworks of entangled relations, in constant mutual transformation, have generated – and continue to generate – new ways of being, for people *and* plants. The concepts of materiality, relationality, meshworks and entanglement allow us to see plants as agents with which we have dynamic relationships. They open up new possibilities to bring together the various strands of archaeology, and to integrate fully archaeological science within theoretical debates. This renewed emphasis on dynamic entanglements allows us to see archaeology again as human ecology (Butzer 1982), but now with people, plants, animals, material culture and environment all receiving equal agency.

Acknowledgements

I would like to thank the Netherlands Institute for Advanced Study for a NIAS Fellowship during which this project was researched. I have benefitted from discussions with Julie-Anne Bouchard-Perron, Chris Gosden, Oliver Harris and Terry Hopkinson, and I am grateful to Richard Bradley, Oliver Harris and Terry Hopkinson for comments on an earlier draft of this paper, and to Eric Cambridge for checking the English text.

References

Bradley, R., 2004. 'Domestication, sedentism, property and time: Materiality and the beginnings of agriculture in Northern Europe', in *Rethinking Materiality: The Engagement of Mind with the Material World*, ed. E. DeMarrais, C. Gosden and C. Renfrew, Cambridge: McDonald Institute for Archaeological Research Monographs, pp. 107–15.

Butzer, K. W., 1982. *Archaeology as Human Ecology*. Cambridge: Cambridge University Press.

Fraser, R. A., Bogaard, A., Schäfer, M., Arbogast, R. and Heaton, T. H. E., 2013. 'Integrating botanical, faunal and human stable carbon and nitrogen isotope values to reconstruct landuse and palaeodiet at LBK Vaihingen an der Erz, Baden-Württemberg', *World Archaeology* 45(3), 492–517.

Fuller, D. Q., Allaby, R. G. and Stevens, C., 2010. 'Domestication as innovation: The entanglement of techniques, technology and chance in the domestication of cereal crops', *World Archaeology* 42(1), 13–28.

Goodman, J., Lovejoy, P. E. and Sherratt, A. (eds), 1995. *Consuming Habits: Drugs in History and Anthropology*. London: Routledge.

Gosden, C., 2004. 'Aesthetics, intelligence and emotions: Implications for archaeology', in *Rethinking Materiality: The Engagement of Mind with the Material World*, ed. E. DeMarrais, C. Gosden and C. Renfrew. Cambridge: McDonald Institute for Archaeological Research Monographs, pp. 33–40.

Gosden, C., 2005. 'What Do Objects Want?', *Journal of Archaeological Method and Theory* 12(3), 193–211.

Harlan, J. R., 1992. *Crops and Man*. 2nd edn. Madison, WI: American Society of Agronomy and Crop Science Society of America.

Harris, O. J. T., and Sørensen, T. F., 2010. 'Rethinking emotion and material culture', *Archaeological Dialogues* 17(2), 145–63.

Hastorf, C. A., 1998. 'The cultural life of early domestic plant use', *Antiquity* 72, 773–82.

Head, L., Atchison, J. and Gates, A., 2012. *Ingrained. A Human Bio-Geography of Wheat*. Farnham: Ashgate.

Hobhouse, H., 1992. *Seeds of Change: Five Plants that Transformed Mankind*. London: Papermac.

Hodder, I., 2011. 'Human-thing entanglement: Towards an integrated archaeological perspective', *Journal of the Royal Anthropological Institute* (N.S.) 17, 154–77.

Ingold, T., 1993. 'The Temporality of Landscape', *World Archaeology* 25(2), 152–74.

Ingold, T., 1996. 'Growing plants and raising animals: An anthropological perspective on domestication', in Harris, D. R. (ed.), *The Origins and Spread of Agriculture and Pastoralism in Eurasia*. London: UCL Press, 12–24.

Ingold, T., 2007. 'Materials Against Materiality', *Archaeological Dialogues* 14(1), 1–16.

Jones, G., Charles, M., Jones, M. K., Colledge, S., Leigh, F. J., Lister, D. A., Smith, L. M. J., Powell, W., Brown, T. A. and Jones, H., 2013. 'DNA evidence for multiple introductions of barley into Europe following dispersed domestications in Western Asia', *Antiquity* 87, 701–13.

Jones, G., Jones, H., Charles, M. P., Jones, M. K., Colledge, S., Leigh, F. J., Lister, D. A., Smith, L. M. J., Powell, W. and Brown, T. A., 2012. 'Phylogeographical analysis of barley DNA as evidence for the spread of neolithic agriculture through Europe', *Journal of Archaeological Science* 39, 3230–8.

Jones, O. and Cloke, P., 2008. 'Non-human agencies: Trees in place and time', in *Material Agency: Towards a Non-Anthropocentric Approach*, ed. C. Knappett and L. Malafouris, New York, NY: Springer, pp. 79–96.

Knappett, C., 2011. *An Archaeology of Interaction. Network Perspectives on Material Culture and Society*. Oxford: Oxford University Press.

Knappett, C. and Malafouris, L. (eds), 2008. *Material Agency: Towards a Non- Anthropocentric Approach*. New York, NY: Springer.

Leach, H. M., 1997. 'The terminology of agricultural origins and food production systems: a horticultural perspective', *Antiquity* 71, 135–48.

Lewis-Williams, D., 2004. *The Mind in the Cave*. London: Thames & Hudson.

Mabey, R., 2010. *Weeds. How Vagabond Plants Gatecrashed Civilisation and Changed the Way We Think about Nature*. London: Profile Books.

Marx, K. and Engels, F., 1966 [1844]. 'Chapter 1: The German Ideology', in *Selected Works*, vol. 1, Moscow: Progress, pp. 16–80.

Mintz, S. W., 1985. *Sweetness and Power: The Place of Sugar in Modern History*. New York, NY: Viking.

Mitchell, W. J. T., 1996. 'What do pictures really want?', *October* 77, 71–82.

Overton, N. J. and Hamilakis, Y., 2013. 'A manifesto for a social zooarchaeology. Swans and other beings in the Mesolithic', *Archaeological Dialogues* 20(2), 111–36.

Pollan, M., 1989. 'Weeds Are Us', *The New York Times Magazine*, 5 November. [online] Available at: http://www.nytimes.com/1989/11/05/magazine/weeds-are-us.html?module=Search&mabReward=relbias%3As. Accessed 26 September 2013.

Pollan, M., 2001. *The Botany of Desire: A Plant's Eye View of the World*. New York, NY: Random House.

Robb, J., 2010. 'Beyond agency', *World Archaeology* 42(4), 493–520.

Robb, J., 2013. 'Material culture, landscape of action and emergent causation. A new model for the origins of the European Neolithic', *Current Anthropology* 54(6), 657–83.

Robb, J. and Harris, O. J. T., 2013. *The Body in History. Europe from the Palaeolithic to the Future*. Cambridge: Cambridge University Press.

Sherratt, A., 1981. 'Plough and pastoralism: Aspects of the secondary products revolution', in I. Hodder, G. Isaac and H. Hammond (eds), *Patterns of the Past: Studies in Honour of David Clarke*. Cambridge: Cambridge University Press, pp. 261–305.

Sofaer, J. R., 2006. *The Body as Material Culture. A Theoretical Osteoarchaeology*. Cambridge: Cambridge University Press.

Tarlow, S., 2000. 'Emotion in archaeology', *Current Anthropology* 41(5), 713–45.

Van der Veen, M., 2008. 'Food as embodied material culture: Diversity and change in plant food consumption in Roman Britain', *Journal of Roman Archaeology* 21, 83–110.

Van der Veen, M., 2011. *Consumption, Trade and Innovation. Exploring the Botanical Remains from the Roman and Islamic Ports at Quseir al-Qadim, Egypt.* Frankfurt: Africa Magna Verlag.

Viola, H. J., 1991. 'Seeds of change', in H. J. Viola and C. Margolis (eds), *Seeds of Change: A Quincentennial Commemoration.* Washington, DC: Smithsonian Institution Press, pp. 11–15.

Wallace, M., Jones, G., Charles, M., Fraser, R., Halstead, P., Heaton, T. H. E. and Bogaard, A., 2013. 'Stable Carbon Isotope Analysis as a Direct Means of Inferring Crop Water Status and Water Management Practices', *World Archaeology* 45(3), 388–409.

Watson, A. M., 1983. *Agricultural Innovation in the Early Islamic World.* Cambridge: Cambridge University Press.

Notes

1. This chapter was first published in 2014 in *World Archaeology*, Debates in World Archaeology, volume 46(2).
2. Throughout the paper where I refer to the 'desires' of plants, this is to be taken metaphorically. It should not be taken to mean that plants, and the history of plant behaviour, have any kind of plant intentionality; that is, no teleological position is implied.
3. This concept has a long pedigree, see, for example, Marx and Engels 1966, 20.

3 PLANTS AS MEDICINE IN THE ANTHROPOCENE

Sarah E. Edwards

Introduction

In this chapter I will evaluate our relationships with plants as medicine in the context of the Anthropocene. Drawing on personal experience working both with scientists and Indigenous/First Nations or local peoples, I will discuss how culture mediates these relationships – and the disparity between different ontological positions, highlighting the social, health and environmental justice issues that these differences raise.

Humanity's interaction with and dependency on plants goes back to the beginning of our existence. Fundamentally, plants supply us with everything that we need for our survival: oxygen to breathe, food to nourish us, materials to make clothes and shelter with, and medicine to treat or prevent sickness and disease – the focus of this chapter.

Prehistoric use of medicinal plants

Evidence suggests that plants have been utilised by different peoples for their medicinal properties since prehistory. For example, archaeobotanical studies show that humans have been using *Cannabis sativa* L. for at least 10,000 years. Cannabis fruit and seed macrofossils were found attached to potsherds in a Mesolithic Age archaeological site in Okinoshima in Japan that were dated to about 10,000 years ago. Early written records exist that describe the medicinal use of the plant, including in ancient Vedic texts from about 800 BCE and in the first known Pharmacopoeia from China, '*Shen Nung Pen Ts'ao Ching*' dating to the first century BCE. Its medical and religious importance to people led to the plant being widely distributed across the globe (Pisanti and Bifulco 2019). Today cannabis and its derivatives are the basis of licensed pharmaceutical medicines used in the treatment of multiple sclerosis spasticity and epileptic seizures (Freeman et al. 2019).

Another notable plant medicine that has been used since prehistory is opium poppy, *Papaver somniferum* L. Archaeological evidence shows that the poppy was used widely for cult ritual or healing purposes in the East Mediterranean dating back at least since the fifth century BCE. A number of Minoan period artefacts depicting poppies have been uncovered on Crete, including the well-known 'poppy goddess', a terracotta figurine from the Late Minoan period, about 1300–1250 BCE, which wears a headdress of three movable hairpins, shaped as slit poppy capsules. The vertical slits on the capsules indicate that the Cretans were aware of the method of extracting opium, which is by incising unripe seed capsules to release the latex (Askitopoulou et al. 2002). Natural derivatives of opium, including morphine and codeine, are used in biomedicine today as analgesics and in anaesthesia (Schiff 2002).

Pharmaceutical industry and use of medicinal plants

The continuing importance of plants to biomedicine is unquestionable: a review of the source of all approved pharmaceutical drugs between 1 January 1981 and 30 September 2019 found that almost a quarter of these are natural products or natural product derived. Plants are of particular relevance to oncology, with over half of cancer drugs derived from natural compounds (Newman and Cragg 2020). These include vinca alkaloids from the Madagascar periwinkle, *Catharanthus roseus* (L.) G.Don used in chemotherapy drugs against testicular carcinoma, leukaemia and both Hodgkin and non-Hodgkin lymphoma (Moudi et al. 2013). Another natural product, paclitaxel, originally discovered from the Pacific yew, *Taxus brevifolia* Nutt., is used in the treatment of lung, breast and ovarian cancers. The vinca alkaloids and paclitaxel share a mechanism of action in their anticancer effects: they bind to the protein tubulin, which prevents the formation of microtubules thereby inhibiting division of cancer cells (van Vuuren et al. 2015). The complexity of these molecules means that it is unlikely that they would have been discovered without natural product research (Howes et al. 2020).

Of the 347,298 known vascular plant species (Simmonds et al. 2020, 546–56), at least 33,443 – or about one in ten – are recorded as having a medicinal use (MPNS 2021). However, only some have been investigated chemically, and it is likely that many new drugs are waiting to be discovered (Howes et al. 2020). In 2015, Youyou Tu from the Institute of Chinese Materia Medica was awarded the Nobel Prize in Medicine for her key contribution to the discovery of a novel therapy against malaria. Tu had investigated ancient Chinese medical texts to uncover the antimalarial properties of *Qinghao* (sweet wormwood), *Artemisia annua* L., isolating the active molecule, artemisinin. Since its discovery, drugs based on artemisinin have made a significant contribution to global health. In addition to its use against malaria, artemisinin is showing promise in other areas, notably as a potential role in cancer treatment (Su and Miller 2015; Wang et al. 2019).

Traditional and herbal medicine

Plants, in the form of herbal medicines, are also important in Traditional Medicine (TM), which plays a significant role in helping to meet primary health care needs in many developing countries. This is especially true in rural communities that lack access to pharmaceutical medicines due to their prohibitive costs, where TM is perceived as being more affordable (Oyebode et al. 2016). According to Medecins Sans Frontieres (MSF), more than one third of the world's population lacks access to essential[1] pharmaceutical medicines, and in the most impoverished areas of Africa and Asia this figure rises to half of the population (MSF 2021).

In addition to their use in TM in developing countries, herbal medicines are increasingly popular in industrialised nations. While exact figures are unknown, it is clear that the value of herbal medicines in international trade is increasing: in 2003 the global value of herbal medicines was estimated at US$60 billion, and by 2012 the value of Traditional Chinese Medicine (TCM) alone was estimated at US$83 billion (Allkin et al. 2017).

Where herbal medicines are not part of a country's own tradition, or integrated into its dominant healthcare system, they are considered to belong to 'Complementary and Alternative Medicine' (CAM). CAM is used to refer to any systems, treatments or therapies that do not fall under accepted 'conventional' medicine – a sociocultural concept, since notions of conventionality change over time and differ between cultures (Edwards et al. 2012). Increasingly with globalisation, there is a pluralistic approach to medicine, for example TCM is gaining in popularity in East Africa (Hsu 2007). However, this is not a new phenomenon: the search for novel commodities, including exotic medicinal plants, was one of the drivers for European colonialism, leading to exploitation of Indigenous peoples and their knowledge (Voeks and Greene 2018). Many of these medicinal plants were brought back from the so-called 'New World', along with associated medicinal knowledge, to be integrated into European pharmacopoeias (Dutfield 2017).

A criticism often levelled at herbal medicines is that they lack scientific evidence. Compared to pharmaceutical medicines there is a lack of high quality randomised controlled trials, in part due to prohibitive costs. However, 'comparative effectiveness' research can be used to assess their effectiveness in everyday practice settings, by comparing two or more health interventions to determine the preferred option for specific patients (Edwards et al. 2015, 9).

When assessing the clinical effectiveness of herbal remedies, scientists rarely consider the non-pharmacological functions of these medicines, which can affect the patient through physiological or psychological means by eliciting a '*meaning response*' (Moerman and Jonas 2002). Traditional herbal remedies are often administered within religious or mythical traditions which help evoke associated meaning, and in turn may result in the patient producing endogenous compounds. The interaction of the patient with a healer may also induce a meaning response, increasing any therapeutic effects of the plant extract itself (Edwards et al. 2015, 9).

Doutor Raiz

> I make prescriptions, but God kills.
> (José Luis '*Dr Raiz*', Fortaleza in Edwards 1998)

Among the *curandeiros* (traditional healers) of the *Nordeste* (Northeast) of Brazil are some with specialist knowledge of plant medicines, who will make up herbal preparations for their patients. The healers commonly employ prayer, in conjunction with the use of herbal medicines, enhancing any meaning response that the remedies elicit in the patient. Those traditional healers who find acclaim as professional herbalists are given the name *Doutor Raiz* ('Doctor Root').

During fieldwork in Brazil in 1998, one such *Dr Raiz*, José Luis, was encountered selling his medicinal preparations from a market stall in Fortaleza, a coastal city in the state of Ceará. Of part native descent, Dr Raiz was born in the interior of Ceará and worked with medicinal plants as a family tradition. Dr Raiz had ten men working for him full time, collecting plants from the state interior, the Amazon and other areas of Brazil. He had been treating patients with plant medicines for over thirty years, and his renown had spread widely, so that people travelled great distances in order to consult him. Patients often returned later with offerings of thanks for his 'miraculous cures'. His healing practice involved a strong metaphysical element as, in addition to his extensive plant knowledge, he was a spirit medium who carried out exorcisms for people affected by malevolent spirits.

Dr Raiz's specialist medicinal plant knowledge had been passed down through an oral tradition, and expanded through his years of experience, along with textbooks, including one by Professor Matos from the Natural Products Laboratory at the Universidade Federal do Ceará (UFC). Traditional knowledge, like science, is dynamic and evolves, and the fact that Dr Raiz used a book based on scientific 'rationality' of the local herbal pharmacopoeia, highlights the interplay between the two forms of knowledge. The most common ailment which Dr Raiz treated with both plant preparations and prayer was *nervoso* ('nerves'), a locally recognised condition associated with both psychological and physical symptoms (Edwards 1998, 29–31).

Farmácias Vivas (Living Pharmacies)

Professor F.J.A. Matos (1924–2008) was an ethnobotanist and phytochemist who initiated the *Projeto Farmácias Vivas do Ceará* (Living Pharmacies Project of Ceará) in 1983 with the aim to provide safe, effective and affordable herbal medicines to local people. As part of the project, a medicinal plant garden was established at UFC, one of the only germplasm banks of medicinal plants in Brazil. Matos also developed a database, a repository of scientific studies on regional medicinal plants (Bonfim et al. 2018).

Knowledge of herbal medicines in the *Nordeste* is held in both the popular and scientific domains, although the two do not always correlate. The plants grown in the Living Pharmacies have been tested in the laboratory for their clinical effectiveness and toxicity, and only species deemed safe are encouraged for use. In contrast, a survey of medicinal plant market stalls in Recife (in the state of Pernambuco) and Fortaleza in 1998 found a number of highly toxic and potentially lethal plants could readily be purchased – although vendors did stress their danger, which was not unknown. Matos expressed his dismay that despite raising awareness of the risks of these plants, people continued to use them.

The majority of medicinal plants sold in the markets were of European or West African origin, reflecting local people's ethnic heritage, a legacy of colonialism and the slave trade. Along with knowledge of specific plants, medical and health beliefs that local people held also reflected this heritage. For example, the use of emetics and purgatives (often toxic plants) was common, due to the widely held belief that sickness must be purged from the body. Concepts based on ancient humoral theory, with health considered a balance between opposites such as hot-cold and wet-dry, as well as a belief in the 'Doctrine of Signatures'[2] were also found. Plants to protect from *mal olhado* (evil eye) were popular, highlighting the non-pharmacological, but symbolic value of medicinal plants in helping disorders of a psychological or emotional basis. Some of the highly toxic plants available in the markets of the *Nordeste* were used as emmenagogues, to regulate 'delayed menstruation', a cognitively ambiguous and euphemistic explanation, providing women with reproductive control (Edwards 1998, 37–8). These local understandings of medicinal plant use provide an explanation as to why local people continued to self-administer toxic plants purchased from the markets, despite warnings of their risks from the scientists at UFC.

Today, state government departments of Ceará run an 'official' medicinal plant garden, and by 2017, 42 Living Pharmacies were active in the state (Bonfim et al. 2018). Matos' vision of the Living Pharmacy project in Ceará has become a template for other states in Brazil and is now a national social health project. As acknowledgment and homage to him, 21 May, his birthday, is known in Brazil as Medicinal Plant Day (Silveira 2008).

Biocultural diversity: Loss of languages and medicinal plant knowledge

In recent years there has been growing awareness of the 'inextricable link' between biological diversity, languages and cultural diversity: studies have shown that there is a strong correlation between languages and 'hotspots' of biological diversity (Maffi 2010; Gorenflo et al. 2011). There are parallel threats to both biodiversity and languages, with predictions that 50–90 per cent of languages will go extinct by the end of this century. Many of these languages are endemic to a particular locality and spoken by relatively few people, indicating their vulnerability. Possible reasons for the co-occurrence of languages and biodiversity are complex, dependent on specific local features, yet given the strong correlation, a functional connection between the two is likely (Gorenflo et al. 2011). The most vulnerable languages are spoken by Indigenous peoples, who although only represent about 5 per cent of the world's human population, are stewards of 80 per cent of Earth's biodiversity (Ogar et al. 2020).

A large proportion of existing medicinal knowledge has been shown to be linked to threatened languages. In a regional study focusing on the Amazon, New Guinea and North America, it was found that 75 per cent of medicinal plant uses are known in only one language. In Indigenous groups, intergenerational knowledge is traditionally transmitted orally. When a language disappears, so does its embodied cultural wisdom, including medicinal plant knowledge (Cámara-Leret and Bascompte 2021).

Wik, Wik Waya and Kugu ethnobiology project

> my parents taught me the name of every tree, every plant, every fish….in twenty years this will all be forgotten. Young people today prefer to live in the busy world.
>
> (Wik-Alkan Traditional Owner 2002, quoted in Edwards 2006)

Before the arrival of Europeans, more than 250 languages with numerous dialects were spoken on the continent of Australia. Today, less than half of these languages are still spoken, and 90 per cent are considered endangered (Karidakis and Kelly 2018; AITSIS 2021a). Concerns regarding the loss of traditional ecological knowledge (TEK) in the First Nation community of Aurukun,[3] as expressed in the Traditional Owner's lament above, led to an invitation in 2001 from Aurukun Shire Council to collaborate on an Ethnobiology project and develop an associated database for the community.

Loss of TEK and associated knowledge of bush medicine was evident in Aurukun, as the impact of colonisation, including historic government policies of segregation and assimilation (Australian Human Rights Commission 1997) continue to be felt. During my fieldwork with First Nation communities in northern Australia I heard first-hand testimonies from mothers who had their children forcibly removed by the authorities. The so-called 'Stolen Generations' were in many cases never seen by their families again. While the colonial legal fiction of *terra nullius*[4] was finally overturned in 1992 following the Mabo native title case (AITSIS 2021b), today European colonialists are replaced by transnational corporations and the globalised economy. These undermine Indigenous rights and pose a threat to their lands, including the non-human beings who inhabit them. Across the globe Indigenous territories are erroneously considered 'wilderness', waiting to be 'developed' (i.e., exploited) to appease the hunger of avaricious marketeers.

Negotiations are currently underway between Wik Waya Traditional Owners and Glencore/Mitsubushi, who hold the lease to develop a bauxite mine in Aurukun Shire (Glencore 2021). A few Wik Traditional Owners I worked with in the early 2000s expressed their concern about the proposed mine, as health and well-being are considered by the Wik to be interlinked with country. I was told that *'people won't be able to live on country anymore and will get sick'*.

Environmental degradation, combined with socio-political factors, have contributed to the loss of Wik TEK and associated practices. These factors include the shift from subsistence economies to one predominantly based on 'passive welfare' – referred to by local people as *'sit-down money'*. Another factor is the disruption to Wik pedagogy from the cessation of sacred schooling decades earlier, when young Wik men reaching adulthood would be segregated from the community to be instructed in Aboriginal Law and practical aspects of caring for country (Edwards and Heinrich 2006).

Like many other Indigenous peoples across the world, the Wik people of Aurukun perceive themselves as part of an extended kinship network, related to all other living beings. This has been termed 'kincentric ecology' by the Native American Ethnobotanist Enrique Salmón (2000). The complex Australian First Nation peoples' totemic and kinship systems are based on reciprocity, with a duty to 'Care for Country', which in turn will care for them. The Wik consider themselves to belong to their 'country' (clan estate), where their spirit emanates from and where it returns to on death. Caring for country, including the plants and animals that live there, therefore shows respect to one's ancestors, as well as ensuring country can care for future generations. The sentience and personhood of other living beings is acknowledged. The lived experience is of 'culture' being within nature, not separate from it.

Opar' (Bush Medicine)

> Young people today do not know much bush medicine – they forget culture, but old people still have culture.
>
> (Wik Aboriginal Health Worker 2003, quoted in Edwards 2006)

Opar' is the Wik-Mungkan[5] word for medicine and may be used to refer to both bush medicine and clinic prescribed pharmaceuticals. However, what constitutes 'medicine' to the Wik differs from the biomedical definition, encompassing a range of holistic treatments to treat disease and promote health and well-being. Sickness and health in Wik terms applies not just to the body, but also one's spirit, society, environment, and country. In addition to medicine to treat the body, there is *opar'* to control the rain and provide protection from lightning strikes, spirits, or the Rainbow Serpent (a significant Aboriginal creator deity).

During fieldwork in Aurukun, I worked closely with the highly regarded *Songman*[6] and principal knowledge custodian, Joe Ngallemetta, who adopted me as his own daughter. Joe was the last person alive in the community who had been brought up traditionally in the bush, being taken into the then Mission as a 14-year-old boy, when his parents exchanged him for some flour, sugar and tobacco with the missionaries. He was also the last native speaker of Kugu-Uwanh, the language lost to the world on his passing – just six months after my return to the UK from the field.

Knowledge of *opar'* was believed to be declining as people lost touch with their clan estates – although the more elderly Wik, especially those living on outstations, used traditional *opar'* in preference to clinic medicine. Prior to mission days, the Wik lived in small family groups and knowledge was passed down the generations in each family and rarely divulged to non-family members. When someone from a different family married into a group, they brought their own knowledge of *opar'* with them. With the establishment of the Mission, this all changed as different clan and family groups were brought into close proximity and more knowledge was exchanged. The Wik considered there was still a clear distinction, however, between *opar'* from clan groups belonging to the eucalypt forest and those belonging to coastal areas, reflecting the different habitats and plant species found growing in them.

Mnemonic devices, similar to the Doctrine of Signatures, whereby sensory cues are used to indicate a particular medicinal use of a plant, facilitate transmission of oral knowledge. For example, a red inner tree bark is used to treat red open sores, and a milky sap is used to promote lactation in breastfeeding mothers. The majority of Wik medicinal plants are used externally or burned to release a smoke to clear negative spirits, like Native American smudging. The plants are considered to have spiritual healing capabilities, in addition to any perceived physiological actions.

Several traditional food plants ('bush tucker') also have medicinal properties, for example eating local wild yams (*Dioscorea spp.*) was considered by the Wik to be beneficial for those with diabetes, through helping to balance blood sugar levels. Scientific evidence supports this observation, as studies on *Dioscorea* species have demonstrated anti-diabetic properties and hypoglycaemic activity of polysaccharides found within the tubers (Li et al. 2017; Padhan and Panda 2020). It is unlikely that there would have been many cases of diabetes in the community prior to the mission days, and it certainly would not have been conveyed of in current biomedical terms, indicating the dynamic nature of TEK. Globally, Indigenous populations are disproportionately affected by type II diabetes and related complications, the consequence of the nutrition transition from traditional foods to a diet high in refined carbohydrates and contributing to the large disparities in life expectancy between Indigenous and non-Indigenous Australians (Giovannini et al. 2016; Nguyen et al. 2016).

As a Wik Traditional Owner stated:

> May [plant] food make you healthy. [In the] Old days people lived to older age – not like now, people dying before their time.
>
> (2001, quoted in Edwards 2006)

The Anthropocene and global demand for medicinal plants

Humanity's impact on planet Earth has been so profound that Earth scientists have proposed a new geological epoch – the Anthropocene, which some argue began in the mid-twentieth century (Syvitski et al. 2020). Scientists belonging to the Intergovernmental Panel on Climate Changes (IPCC) have reported changes in climate unprecedented in thousands or hundreds of thousands of years, with rise in sea levels and temperature expected to increase by at least 1.5°C in the next two decades (IPCC 2021). Simultaneously we are witnessing extinction rates about 1,000 times greater than expected background rates (Pimm et al. 2014).

According to Kew's *State of the World's Plants and Fungi 2020*, 39.4 per cent, roughly two in five plants, are estimated to be threatened with extinction (Antonelli et al. 2020). Habitat loss due to changes in land use (including from agriculture, mining and urbanisation), is one of the main drivers of loss of plant species, followed by biological resource use (Nic Lughada et al. 2020). The global demand for medicinal species leads to overharvesting, threatening wild populations (Allkin et al. 2017). As many as 60–90 per cent of all medicinal and aromatic plants in trade are currently wild harvested. However, the conservation status is unknown for 93 per cent of these plants. The volume of trade is high, and growing, with a threefold increase since 1999: 1.3 billion kg of botanical ingredients were exported by China alone in 2013. The increased demand for wild plant ingredients leads to traditional sustainable harvesting practices being replaced by more intensive and environmentally harmful extraction methods (Jenkins et al. 2018).

Wild harvested medicinal plants are generally considered by consumers to be superior to cultivated plants, due to the belief in their greater effectiveness in treating conditions. Wild ginseng roots, for example, are five to ten times more valuable than those produced using artificial propagation methods (Schippmann et al. 2002). This may be due to the presence of secondary metabolites resulting from growing conditions in the wild, including mycorrhizal fungi in the soil and/or competition with other species (Schippmann et al. 2006). When ex-colleagues from Kew worked on a project in West Africa with traditional healers to establish a medicinal plant garden (to reduce unsustainable wild harvesting), the traditional healers stressed the importance of transplanting soil along with the plants to the new garden. They explained to the Kew scientists that spirits in the soil were needed for the plants to grow and be effective. Scientists are likely to dismiss mention of spirits as irrational superstition, however, in this case, the scientists were aware of the likelihood of living entities in the soil: microscopic mycorrhizal fungi that were invisible to the naked eye. Both healers and scientists were conveying different interpretations of the same reality.

Sustainability, social justice and safety

Over-exploitation of medicinal plants is compounded by habitat loss, environmental degradation, and the low wages that many harvesters receive. The income generated by harvesting medicinal plants may be the only source of livelihood for some rural populations, who generally receive a minimum of the final retail revenue of the plant material (Booker et al. 2012; Volenzo and Odiyo 2020). However, as a plant becomes rarer, its commercial value increases – creating a positive feedback loop resulting in an increased threat to the species. The scarcity of particular medicinal plants means that often they are substituted, deliberately or inadvertently, with a different species – leading to potential safety issues for the patient (Allkin et al. 2017).

Commodification of the sacred

Historically, colonial powers exploited TEK without recourse, treating the 'discovery' of a plant's medicinal properties as a new scientific development, without acknowledging any intellectual input of the TEK holders. Colonisers and corporations used intellectual property laws derived from Eurocentric conceptualisation of ownership, knowledge, authorship and property to maximise their profits, patenting drugs developed through bioprospecting (Anderson 2015).

Today, 6.2 per cent of plant species are associated with patents, which include 7,595 patents in medicine and 2,941 patents in cosmetics. Scientists have advocated for 'nature-based solutions', using the patent system to commercialise novel plant-based products, suggesting this may lift TEK holders out of poverty and incentivise the conservation of biodiversity (Simmonds et al. 2020). However, this demonstrates how science does not exist in a vacuum but is shaped according to prevailing socio-political narratives. Neoliberal ideologies have increasingly dominated biological conservation discourse, promoting economic and utilitarian models of the natural world, and conservationists have adopted the language of economists, such as 'natural capital', 'valuing nature', etc. (Guerry et al. 2015). These models create a relationship with nature that is unilateral rather than reciprocal, exploitative rather than nurturing, and undermines traditional holistic concepts and ontologies, including of the sacredness of plants (Posey 2002).

Agency of plants and healing the land

What of the plants themselves?

While recognition of the sentience and agency of non-human beings is rooted in many Indigenous peoples' relationships with the natural world (e.g. Baker 2021; Wall Kimmerer 2013), it is only in recent years that multi-species ethnography has evolved as a new field of study in anthropology (Kirksey and Helmreich 2010). Other studies, including the work of Suzanne Simard (2018) and those presented by Jeremy Narby (2005), suggest an innate intelligence of plants, highlighting their communication, memory, kin recognition and interspecies cooperation – indicative of more than simply reactive responses to environmental triggers.

A multi-species ethnographic approach was used in a recent British Ecological Society funded project, working collaboratively as an ethnobotanist with an artist, Ami Marsden, and farmers, Kate and Illtud Dunsford, in Carmarthenshire, west Wales. In this study, the landscape was investigated to reveal the narratives of the plants encountered. Interacting with these plants, and listening to their stories, allowed a re-evaluation of people-plant relationships and the role of plants as healers, not just of people – but as transformative agents of the land itself.

The study site, Felin y Glyn Farm, is a mixed farm of 167 acres in the Gwendraeth Valley near Pontyates, where Illtud's family have farmed for over 300 years (*Coed Gwenllian* 2021). Illtud and Kate view themselves as guardians or stewards of the land – a view also frequently held by Indigenous and traditional peoples (Posey 2002) – aware of their temporality in a landscape rich in history and folklore. The land was historically part of a great forest, Coed y Glyn (or Glyn Forest), which stretched from the coast inland to the Carmarthenshire Fens, and is mentioned in the Mabinogion, a collection of stories compiled in Middle Welsh in the twelfth to thirteenth centuries (*Coed Gwenllian* 2021). The area was also once part of a coal and associated ironworks industry, with anthracite being mined in the Gwendraeth Valley colliery in the nineteenth century, until its demise in the early twentieth century (Welsh Coal Mines 2021).

Beside a stream on the farm remnants of structures from old mine workings are hidden beneath vegetation, which has regenerated and grown over them to form a mixed deciduous wooded area – now a peaceful sanctuary where Kate and Illtud often retreated to during the Covid-19 lockdown. Studies have demonstrated empirical evidence that exposure to forests and trees benefits human health, including by reducing stress and enhancing immune system function (FAO 2020). Mindfulness and the Japanese practice of *Shinrin-Yoku* ('forest-bathing'), has been proposed as a cost-effective modality to alleviate mental health issues related to the Covid-19 pandemic and beyond (Timko Olson et al. 2020).

As we walked across the farm meadows, plants metaphorically called out and attracted our attention. One of these was evening primrose, *Oenothera biennis* L., its large yellow, and according to Kate and Illtud, rather '*blousy*' flower seeming to be out of place. Interwoven with this plant is a story of colonialism and exploitation, having been taken to Britain, where it has naturalised, from North America during the 1600s. The knowledge of the evening primrose's use as a healing herb was held by various Native American tribes, including the Cherokee, Iroquois, Ojibwe and Potawatomi, who used the plant to treat skin complaints, haemorrhoids, for strength and as a stimulant amongst other uses (Moerman 2003). In seventeenth-century Europe evening primrose became a popular folk remedy, and was called 'King's cure-all' (NCCIH 2020). Today evening primrose is cultivated for its oil, which is extracted from the seeds and sold in Britain and elsewhere for self-administered treatment of pre-menstrual syndrome, menopausal symptoms, rheumatoid arthritis, dermatitis and eczema (Edwards et al. 2015, 144–8).

A few other medicinal plants were found growing in the meadows, especially around the field margins, including St John's Wort, *Hypericum perforatum* L., which flowers in midsummer. Its name marks the time of the feast day of St John the Baptist, 24 June, when its yellow flowers would traditionally be gathered. In antiquity the plant was deemed to have supernatural powers, offering protection from evil spirits; its therapeutic uses were documented by ancient Greek herbalists, including Hippocrates, Dioscorides and Galen (Bilia et al. 2002). Today it is used externally to treat minor wounds, bruises, burns and swellings, and internally to treat low mood, anxiety and depression. There is considerable clinical data to show that it as effective as synthetic antidepressants in treating certain types of depression – although it has a high profile of interaction with many pharmaceutical drugs (Edwards et al. 2015, 335–9). As with evening primrose, the story of St John's Wort is linked to people's migration and the plant has been spread to other parts of the globe outside of its native distribution. It was first recorded in North America in the 1700s, where it is now considered an invasive and 'noxious weed', crowding out native species and forage on pasturelands, as well as being toxic to livestock (USDA 2021).

One tall herbaceous plant that stood out on the farm, growing in thick stands along the old canal that transected the meadows, was Himalayan balsam, *Impatiens glandulifera* Royle – listed as an invasive alien plant of concern by the UK government (DEFRA and APHA 2020). The plant was introduced to Europe in 1839, from seeds sent to Kew by Dr. Royle from Kashmir in the Himalayas. Since then, it has naturalised in the British Isles and is now widespread, growing on the banks of waterways and on waste ground, especially along rivers associated with industry such as in South Wales and the north of England (Beerling and Perrins 1993).

The success of the plant in colonising these post-industrial and often polluted landscapes, in addition to its prolific seed production and dispersal, is in part due to its tolerance to environmental pollutants. Research has shown that *I. glandulifera* hyperaccumulates cadmium, a metal which poses a severe risk to the health of both plants and animals, suggesting that Himalayan balsam can play an important role in phytoremediation of areas contaminated by toxic heavy metals (Coakley et al. 2019). On our walk across the farm, the attractiveness of the Himalayan balsam's rich nectar to local bumblebee species was self-evident, as the flowers hummed with the buzzing of the bees. The plant is used in Bach Flower remedies including for 'extreme mental tension' (PFAF 2021). Bioactive steroidal compounds isolated from *I. glandulifera* have demonstrated anti-cancer potential against hepatocellular carcinoma cells *in vitro*, while other compounds isolated from the plant have shown antidepressant-like effect (Coakley and Petti 2021).

Another invasive non-native plant found growing on the farm is Japanese knotweed, *Reynoutria japonica* Houtt (also known under its synonym *Fallopia japonica*), which by law in the UK must be prevented from spreading into the wild due to the ecological damage it can cause (Environment Agency, DEFRA and Natural England, 2018). Like Himalayan balsam, Japanese knotweed rapidly colonises contaminated riparian habitats and is tolerant of heavy metals, accumulating lead and cadmium in its organs, demonstrating phytoremediation potential (Ibrahimpašić et al. 2020). The plant is native to East Asia, where it is a popular TCM herb, used in the treatment of several conditions, including inflammation, favus, jaundice, hyperlipidaemia and cancer, although its pharmacological mechanisms of action are currently unknown (Peng et al. 2013).

In the spring, stinging nettles, mainly *Urtica dioica* L., are found in abundance on the farm. This plant is generally considered a nuisance ruderal weed in Britain, yet it improves soils over-fertilised with nitrate and phosphate, reduces heavy metal content in contaminated soils and improves local biodiversity (nettles support more than 40 species of insect). Nettles are also useful as a raw material for textiles and have medicinal properties as well as being edible (Kregiel et al. 2018).

Learning of these plants' stories reframed the farmers' relationship with them. Rather than perceiving the plants simply as nuisance weeds, Kate and Illtud began to see the plants' lives as interlinked with human and non-human beings: as medicine not only of people, but as healers of the land, playing an active role in undoing the legacy of environmental harm wrought by the industrial past.

The ability of plants to heal and regenerate damaged ecosystems, if they are allowed to, should not be underestimated. The gorse plant, *Ulex europaeus* L., is considered an invasive non-native plant of pastureland in New Zealand. In a conservation project on Bank's Peninsula, the botanist Hugh Wilson uses a 'minimal interference' approach, allowing the gorse to help restore native forest in the Hinewai Reserve. The gorse provides protective cover for native saplings, as well as fixing nitrogen in the soil, and dies off once the native trees grow and shade it, facilitating succession to native woodland (Hinewai Reserve 2021). Another example demonstrating how environmental damage can be reversed by plants if given the opportunity, is the area surrounding the Chernobyl nuclear plant, which was made into an exclusion zone due to initial high levels of radioactivity, after the worst nuclear accident in human history in 1986. More than 30 years later, the area is now rich in biodiversity, with hundreds of species of plants and animals, including more than 60 rare species. The rich natural forest ecosystem that has developed helps cleanse contaminated land and waterways, is more resilient to climate change and wildfires, and is helping to sequester carbon (UNEP 2020).

Conclusions

Humanity currently faces multiple health and environmental challenges: a pandemic, emerging antibiotic resistance and ecological crises – climate change, the rapid loss of biodiversity and the concomitant loss of Indigenous knowledge systems. More than ever our dependence on medicinal plants is evident, yet many are threatened with extinction. Conservation polices must recognise the importance of biocultural diversity in protecting plants and their associated medicinal knowledge. Further integrative studies are needed that investigate both the sociocultural and pharmacological aspects of medicinal plants in the context of sustainable use and trade. Importantly, the transformative ability of plants in their incredible capacity to restore degraded ecosystems, including their planetary role in sequestering carbon, must be recognised and encouraged. As our Indigenous colleagues have pointed out, plants are our allies.

References

AITSIS, 2021a. *Living languages*. [online] Available at: *https://aiatsis.gov.au/explore/living-languages*. Accessed 10 October 2021.

AITSIS, 2021b. *Overturning the doctrine of Terra Nullius: the Mabo case*. [online] Available at: *https://aiatsis.gov.au/sites/default/files/research_pub/overturning-the-doctrine-of-terra-nullius_0_3.pdf*. Accessed 19 December 2021.

Allkin, R., Patmore, K., Black, N. et al., 2017. 'Useful plants – Medicines: current resource and future potential', in K. Willis (ed.), *State of the World's Plants 2017*. London: Royal Botanic Gardens, Kew, pp. 22–9. [online] Available at: *https://stateoftheworldsplants.org*. Accessed 4 October 2021.

Anderson, J. E., 2015. 'Indigenous knowledge and intellectual property rights', *International Encyclopedia of the Social and Behavioral Sciences* (Second Edition) 11, 769–78, doi: *10.1016/B978-0-08-097086-8.64078-3*.

Antonelli, A., Fry, C., Smith, R. J. et al., 2020. *State of the World's Plants and Fungi 2020*. London: Royal Botanic Gardens, Kew, doi: *10.34885/172*.

Askitopoulou, H., Ramoutsaki, I. A. and Konsolaki, E., 2002. 'Archaeological evidence on the use of opium in the Minoan world', *International Congress Series* 1242, 23–9, doi: *10.1016/S0531-5131(02)00769-0*.

Australian Human Rights Commission, 1997. *Bringing them Home. Report of the National Inquiry into the Separation of Aboriginal and Torres Strait Islander Children from Their Families*, April 1997. [online] Available at: *https://humanrights.gov.au/our-work/bringing-them-home-report-1997*. Accessed 15 December 2021.

Baker, J. M., 2021. 'Do berries listen? Berries as indicators, ancestors, and agents in Canada's Oil Sands Region', *Ethnos* 86(2), 273–94, doi: *10.1080/00141844.2020.1765829*.

Beerling, D. J. and Perrins, J. M., 1993. '*Impatiens glandulifera* Royle (*Impatiens roylei* Walp.)', *Journal of Ecology* 81(2), 367–82, doi: *10.2307/2261507*.

Bilia, A. R., Gallori, S. and Vincieri, F. F., 2002. 'St John's wort and depression: Efficacy, safety and tolerability – an update', *Life Sciences* 70(26), 3077–96, doi: *10.1016/S0024-3205(02)01566-7*.

Bonfim, D. Y. G., Gomes, A. B., Brasil, A. R. L. et al., 2018. 'Diagnóstico situacional das farmácias vivas no estado do Ceará [Situational diagnosis of existing healthy living pharmacies in the state of Ceará]', *Journal of Management and Primary Health Care* 9, e15, doi: *10.14295/ jmphc.v9i0.543.*

Booker, A., Johnston, D. and Heinrich, M., 2012. 'Value chains of herbal medicines – Research needs and key challenges in the context of ethnopharmacology', *Journal of Ethnopharmacology* 140(3), 624–33, doi: *10.1016/j.jep.2012.01.039.*

Borch, M., 2001. 'Rethinking the origins of *terra nullius*', *Australian Historical Studies* 32(117), 222-39, doi: *10.1080/10314610108596162.*

Cámara-Leret, R. and Bascompte, J., 2021. 'Language extinction triggers the loss of unique medicinal knowledge', *PNAS* 118(24), e2103683118, doi: *10.1073/pnas.2103683118.*

Coakley, S., Cahill, G., Enright, A. M. et al., 2019. 'Cadmium hyperaccumulation and translocation in *Impatiens glandulifera*: from foe to friend?', *Sustainability* 11(18), 5018, doi: *10.3390/su11185018.*

Coakley, S. and Petti, C., 2021. 'Impacts of the invasive *Impatiens glandulifera*: lessons learned from one of Europe's top invasive species', *Biology* 10(7), 619, doi: *10.3390/biology10070619.*

Coed Gwenllian: The Cultural Renewal of a Rural Landscape. [online] Available at: *https://gwenllian.co.uk/.* Accessed 10 October 2021.

Court, W. E., 1985. 'The doctrine of signatures or similitudes', *Trends in Pharmacological Sciences* 6, 225–7, doi: *10.1016/0165-6147(85)90104-X.*

Dutfield, G., 2017. 'TK unlimited: The emerging but incoherent international law of traditional knowledge protection', *Journal of World Intellectual Property* 20(5–6), 144–59, doi: *10.1111/jwip.12085.*

DEFRA and APHA, 2020. *Guidance: Invasive non-native (alien) plant species: rules in England and Wales.* [online] Available at: *https:// www.gov.uk/guidance/invasive-non-native-alien-plant-species-rules-in-england-and-wales#list-of-invasive-plant-species.* Accessed 9 October 2021.

Edwards, S. E., 1998. 'Medicinal Plant Sociocultural Concepts and Terms in Northeast Brazil. MSc Medical Anthropology dissertation' (unpublished MSc dissertation, UCL).

Edwards, S. E., 2006. 'Medical Ethnobotany of Wik, Wik-Way and Kugu peoples of Cape York Peninsula, Australia: an integrated collaborative approach to understanding traditional phytotherapeutic knowledge and its application' (unpublished PhD thesis, School of Pharmacy, University of London).

Edwards, S. E. and Heinrich, M., 2006. 'Redressing cultural erosion and ecological decline: The Aurukun Ethnobiology Project', *Environment, Development and Sustainability* 8(4), 569–83.

Edwards, S. E., da Costa Rocha, I., Lawrence, M. J., Cable, C. and Heinrich, M., 2012. 'A reappraisal of herbal medicinal products', *Nursing Times* 108(39), 24–7. PMID: 23155905.

Edwards, S. E., da Costa Rocha, I., Williamson, E. M. and Heinrich, M., 2015. *Phytopharmacy: An Evidence-Based Guide to Herbal Medicinal Products*. London: John Wiley and Sons, Ltd.

Environment Agency, DEFRA, and Natural England, 2018. *Guidance: Prevent Japanese Knotweed from Spreading*. [online] Available at: *https://www.gov.uk/guidance/prevent-japanese-knotweed-from-spreading*. Accessed 10 October 2021.

FAO, 2020. 'Forests for human health and well-being: Strengthening the forest-health-nutrition-nexus', *Forestry Working Paper* No. 18. Rome, *doi: 10.4060/cb1468en*.

Freeman, T. P., Hindocha, C., Green, S. F. and Bloomfield, M. A. P., 2019. 'Medicinal use of cannabis based products and cannabinoids', *British Medical Journal* 365, l1141, *doi: 10.1136/bmj.l1141*.

Giovannini, P., Howes, M. J. R. and Edwards, S. E., 2016. 'Medicinal plants used in the traditional management of diabetes and its sequelae in Central America: A review', *Journal of Ethnopharmacology* 184, 58–71, *doi: 10.1016/j.jep.2016.02.034*.

Glencore, 2021. *Aurukun Bauxite Project*. [online] Available at: *https://www.glencore.com.au/operations-and-projects/aurukun*. Accessed 17 December 2021.

Gorenflo, L. J., Romaine, S., Mittermeier, R. A. and Walker-Painemiller, K., 2011. 'Co-occurrence of linguistic and biological diversity in biodiversity hotspots and high biodiversity wilderness areas', *PNAS* 109(21), 8032–7, *doi: 10.1073/pnas.1117511109*.

Guerry, A. D., Polasky, S., Lubchenco, J. et al., 2015. 'Natural capital and ecosystem services informing decisions: from promise to practice', *PNAS* 112(24), 7348–55. *doi: 10.1073/pnas.1503751112*.

Hinewai Reserve, 2021. *The official website of the Hinewai Reserve*. [online] Available at: *https://www.hinewai.org.nz/*. Accessed 10 October 2021.

Howes, M.-J. R., Quave, C. L., Collemare, J., et al. 2020. 'Molecules from nature: Reconciling biodiversity conservation and global healthcare imperatives for sustainable use of medicinal plants and fungi', *Plants, People, Planet* 2(5), 463–81, doi: 10.1002/ppp3.10138.

Hsu, E., 2007. 'Chinese medicine in East Africa and its effectiveness', *IIAS Newsletter* 45, 22. https://www.iias.asia/sites/default/files/nwl_article/2019-05/IIAS_NL45_22.pdf.

Ibrahimpašić, J., Jogić, V., Toromanović, M. et al., 2020. 'Japanese knotweed (*Reynoutria japonica*) as a phytoremediator of heavy metals', *Journal of Agricultural, Food and Environmental Sciences* 74(2), 45–53.

IPCC, 2021. *Climate change, widespread, rapid and intensifying – IPCC.* [online] Available at: https://www.ipcc.ch/2021/08/09/ar6-wg1-20210809-pr/. Accessed 4 October 2021.

Jenkins, M., Timoshyna, A. and Cornthwaite, M., 2018. *Wild at Home: Exploring the global harvest, trade and use of wild plant ingredients.* TRAFFIC report, June 2018. [online] Available at: https://www.traffic.org/site/assets/files/9241/wild-at-home.pdf. Accessed 4 October 2021.

Karidakis, M. and Kelly, B., 2018. 'Trends in indigenous language use', *Australian Journal of Linguistics* 38(1), 105–26, doi: 10.1080/07268602.2018.1393861.

Kirksey, S. E. and Helmreich, S., 2010. 'The emergence of multispecies ethnography', *Cultural Anthropology* 25(4), 545–76, doi: 10.1111/j.1548-1360.2010.01069.x.

Kregiel, D., Pawlikowska, E. and Antolak, H., 2018. '*Urtica* spp.: Ordinary plants with extraordinary properties', *Molecules* 23(7), 1664, doi: 10.3390/molecules23071664.

Li, Q., Li., W., Gao, Q. et al., 2017. 'Hypoglycemic effect of Chinese yam (*Dioscorea opposita* rhizome) polysaccharide in different structure and molecular weight', *Journal of Food Science* 82(10), 2487–94. doi: 10.1111/1750-3841.13919.

Maffi, L., 2010. 'What is Biocultural Diversity', in L. Maffi and E. Woodley (eds), *Biocultural Diversity Conservation: A Global Sourcebook.* London, Washington DC: Earthscan, pp. 3–11.

Moerman, D. E., 2003. *Native American Ethnobotany Database.* [online] Available at: http://naeb.brit.org/. Accessed 7 October 2021.

Moerman, D. E. and Jonas, W. B., 2002. 'Deconstructing the placebo effect and finding the meaning response', *Annals of Internal Medicine* 136(6), 471-6, doi: 10.7326/0003-4819-136-6-200203190-00011.

Moudi, M., Go, R., Yien, C. Y. S. and Nazre, M., 2013. 'Vinca alkaloids', *International Journal of Preventive Medicine* 4(11), 1231–5. PMID: 24404355.

MPNS, 2021. *Medicinal Plant Names Services Portal v10.* [online] Available at: *https://www.kew.org/science/our-science/science-services/ medicinal-plant-names-services.* Accessed 29 September 2021.

MSF, 2021. *Access to medicines.* [online] Available at: *https://msf.org.uk/ issues/access-medicines.* Accessed 29 September 2021.

Narby, J., 2005. *Intelligence in Nature: An Inquiry into Knowledge.* New York, NY: Jeremy P. Tarcher/Penguin.

NCCIH, 2020. *Evening Primrose Oil.* [online] Available at: *https://www. nccih.nih.gov/health/evening-primrose-oil.* Accessed 7 October 2021.

Newman, D. J. and Cragg, G. M., 2020. 'Natural products as sources of new drugs over the nearly four decades from 01/1981 to 09/2019', *Journal of Natural Products* 83(3), 770-803. doi: *10.1021/acs. jnatprod.9b01285.*

Nguyen, H. D., Chitturi, S. and Maple-Brown, L. J., 2016. 'Management of diabetes in Indigenous communities: Lessons from the Australian Aboriginal population', *Internal Medicine Journal* 46(11), 1252–9, doi: *10.1111/imj.13123.*

Nic Lughada, E., Bachman, S. P., Leão, T. C. C. et al., 2020. 'Extinction risk and threats to plants and fungi', *Plants, People, Planet* 2, 389–408, doi: *10.1002/ppp3.10146.*

Ogar, E., Pecl, G. and Mustonen, T., 2020. 'Science Must Embrace Traditional and Indigenous Knowledge to Solve Our Biodiversity Crisis', *One Earth* 3(2), 162–5. doi: *10.1016/j.oneear.2020.07.006.*

Oyebode, O., Kandala, N. B., Chilton, P. J. et al., 2016. 'Use of traditional medicine in middle-income countries: A WHO-SAGE study' *Health Policy and Planning* 31(8), 984-91, doi: *10.1093/heapol/czw022.*

Padhan, B. and Panda, D., 2020. 'Potential of neglected and underutilized yams (*Dioscorea* spp.) for improving nutritional security and health benefits', *Frontiers in Pharmacology* 11, 496, doi: *10.3389/ fphar.2020.00496.*

Peng, W., Qin, R., Li, X. et al., 2013. 'Botany, phytochemistry, pharmacology, and potential application of *Polygonum cuspidatum* Sieb.et Zucc.: A review', *Journal of Ethnopharmacology* 148(3), 729–45, doi: *10.1016/j.jep.2013.05.007.*

PFAF, 2021. *Plants for a Future Database. Impatiens glandulifera – Royle.* [online] Available at: *https://pfaf.org/user/Plant.aspx?LatinName=Impat iens+glandulifera.* Accessed 10 October 2021.

Pimm, S. L., Jenkins, C. N., Abell, R. et al., 2014. 'The biodiversity of species and their rates of extinction, distribution, and protection', *Science* 344(6187), 1246752, doi: 10.1126/science.1246752.

Pisanti, S. and Bifulco, M., 2019. 'Medical cannabis: A plurimillennial history of an evergreen', *Journal of Cellular Physiology* 234(6), 8342–51. doi: 10.1002/jcp.27725.

Posey, D. A., 2002. 'Commodification of the sacred through intellectual property rights', *Journal of Ethnopharmacology* 83(1–2), doi: 10.1016/s0378-8741(02)00189-7.

Salmón, E., 2000. 'Kincentric ecology: indigenous perceptions of the human-nature relationship', *Ecological Applications*, 10(5), 1327–32, doi: 10.2307/2641288.

Schiff, P. L., 2002. 'Opium and its alkaloids', *American Journal of Pharmaceutical Education* 66, 186–94.

Schippmann, U., Leaman, D. J. and Cunningham, A. B., 2002. 'Impact of cultivation and gathering of medicinal plants on biodiversity: global trends and issues', in *Biodiversity and the Ecosystem Approach in Agriculture, Forestry and Fisheries. Satellite event on the occasion of the Ninth Regular Session of the Commission on Genetic Resources for Food and Agriculture. Rome, 12–13 October 2002*, FAO 2002. Rome: Inter-Departmental Working Group on Biological Diversity for Food and Agriculture. [online] Available at: *https://www.fao.org/3/aa010e/aa010e00.htm*. Accessed 19 December 2021.

Schippmann, U., Leaman, D. and Cunningham, A. B., 2006. 'A comparison of cultivation and wild collection of medicinal and aromatic plants under sustainability aspect', in R. J. Bogers, L. E. Craker and D. Lange (eds), *Medicinal and Aromatic Plants*. Berlin: Springer, pp. 75–95.

Silveira, E. R., 2008. 'Editorial: In memoriam – Francisco José de Abreu Matos – 1924–2008', *Revista Brasileira de Farmacognosia* 18 (suppl), doi: 10.1590/S0102-695X2008000500001.

Simard, S. W., 2018. 'Mycorrhizal networks facilitate tree communication, learning, and memory', in F. Baluska, M. Gagliano and G. Witzany (eds), *Memory and Learning in Plants*. Cham: Springer, pp. 191–213, doi: 10.1007/978-3-319-75596-0_10.

Simmonds, M., Fang, R. Wyatt, L., Bell, E., Alkin, B., Forest, F., Wynberg, R., da Silva, M., Zhang, B. G., Liu, J. S., Qi, Y. D. and Demissew, S., 2020. 'Biodiversity and patents: Overview of plants and fungi covered by patents', *Plants, People, Planet* 2(5), doi: 10.1002/ppp3.10144.

Su, X. and Miller, L. H., 2015. 'The discovery of artemisinin and Nobel Prize in Physiology or Medicine', *Science China Life Sciences* 58(22), pp. 1175–9. doi: 10.1007/s11427-015-4948-7.

Syvitski, J., Waters, C. N., Day, J., Milliman, J. D., Summerhayes, C., Steffen, W., Zalasiewicz, J., Cearreta, A., Gałuszka., A., Hajdas, I., Head, M. J., Leinfelder, R., McNeill, J. R., Poirier, C., Rose, N. L., Shotyk, W., Wagreich, M. and Williams, M., 2020. 'Extraordinary human energy consumption and resultant geological impacts beginning around 1950 CE initiated the proposed Anthropocene Epoch', *Communications Earth and Environment* 1, 32, doi: 10.1038/s43247-020-00029-y.

Timko Olson, E. R., Hansen, M. M. and Vermeesch, A., 2020. 'Mindfulness and Shinrin-yoku: Potential for physiological and psychological interventions during uncertain times', *International Journal of Environmental Research and Public Health* 17(24), 9340, doi: 10.3390/ijerph17249340.

UNEP, 2020. *How Chernobyl has become an unexpected haven for wildlife.* [online] Available at: https://www.unep.org/news-and-stories/story/how-chernobyl-has-become-unexpected-haven-wildlife. Accessed 7 October 2021.

USDA, 2021. *National Invasive Species Information Center – St. Johnswort.* [online] Available at: https://www.invasivespeciesinfo.gov/terrestrial/plants/st-johnswort. Accessed 9 October 2021.

van Vuuren, R. J., Visagie, M. H., Theron, A. E. et al., 2015. 'Antimitotic drugs in the treatment of cancer', *Cancer Chemotherapy and Pharmacology* 76, 1101–12. doi: 10.1007/s00280-015-2903-8.

Voeks, R. and Greene, C., 2018. 'God's healing leaves: The colonial quest for medicinal plants in the Torrid Zone', *Geographical Review* 108(4), 545–65, doi: 10.1111/gere.12291.

Volenzo, T. and Odiyo, J., 2020. 'Integrating endemic medicinal plants into the global value chains: The ecological degradation challenges and opportunities', *Heliyon* 6(9), e04970, doi: 10.1016/j.heliyon.2020.e04970.

Wall Kimmerer, R., 2013. *Braiding Sweetgrass: Indigenous Wisdom, Scientific Knowledge, and the Teachings of Plants.* Minneapolis, MN: Milkweed Editions.

Wang, J., Xu, C., Wong, Y. K. et al., 2019. 'Artemisinin, the magic drug discovered from traditional Chinese medicine', *Engineering*. 5(1), 32–9. doi: 10.1016/j.eng.2018.11.011.

Welsh Coal Mines, 2021. *A Welsh Coal Mines Web Page: Gwendraeth, Carmarthenshire.* [online] Available at: *http://www.welshcoalmines.co.uk/Carm/Gwendraeth.htm*. Accessed 7 October 2021.

WHO, 2021. *WHO model list of essential medicines – 22nd list, 2021.* [online] Available at: *https://www.who.int/publications/i/item/WHO-MHP-HPS-EML-2021.02*. Accessed 8 December 2021.

Notes

1. 'Essential' medicines are those that satisfy the priority health care needs of a population. The WHO maintains a model list of essential medicines (WHO 2021).
2. The Doctrine of Signatures or Similitudes states that 'like cures like' and each medicinal substance from nature has a feature that indicates the disease for which it is a remedy or resembles the body part requiring treatment. The Doctrine was promoted by the Swiss physician Paracelsus in the sixteenth century, but developed independently centuries earlier in disparate areas of the world (Court 1985).
3. Aurukun is one of the largest Aboriginal communities on Cape York Peninsula in Far North Queensland, with a population of about 1,200 people. The Wik, Wik Waya and Kugu peoples of Aurukun (collectively known as Wik) belong to five spirit clan groups and 15 language groups.
4. *Terra nullius* – means 'land belonging to no one'. In 1770 Captain Cook proclaimed the coast of New Holland (Australia) a British territory. Despite awareness of the existence of Indigenous inhabitants, the land was considered *terra nullius* – a European legal term that allowed European colonial powers to take control of unclaimed territory (Borch 2001).
5. Wik-Mungkan is the *lingua franca* of Aurukun.
6. *Songmen* are special performers who learn the sacred songs and dances passed down from the Ancestors which play a central role in ceremonies. Totemic creation beings sung the landforms, plants and animals into existence during the Dreaming.

4 THE WORLD TREE
Humans, Trees and Creation on the Sierra Nevada de Santa Marta

Falk Parra Witte

Introduction

Although plants seem to occupy a central role in the ontological patterns of Chibcha Peoples in Colombia and Central America (Niño Vargas 2020), due to historical ruptures, contrasting environments, and diverse lifestyles these patterns are not easy to establish (Halbmayer 2020). This chapter aims to help fill these gaps by giving an insight into the understanding of, and relation to, trees of a Chibcha-speaking Indigenous group from the 'Isthmo-Colombian Area'. There, the topic of plants has been less studied than in Amerindian regions more dominant in the anthropological literature such as Amazonia or the Andes. In addition, the present study takes up Rival's (1998, 1) call to pay greater attention to the social role of trees in people's lives, given how trees feature less prominently in anthropological writing than other plants and animals.

During a period of anthropological fieldwork[1] among the Kogi of the Sierra Nevada de Santa Marta, an isolated mountain massif in northern Colombia shared with the Ika, Wiwa and Kankuamo Peoples, everything I discovered about trees was part of a wider immersion into the Kogi understanding of, and relation to, the world (Parra Witte 2018). While I experienced the practical manifestation of this ontology in daily Kogi life, the core of my research was a structured learning course that I underwent with Kogi spiritual leaders and knowledge experts, the Mamas. 'Trees' were one of four main topics that we covered. It followed those of 'Water' and 'Earth', where I was taught about the spiritual origins of existence and how the world then took shape and materialised, and came before 'Food', which laid out the ecological relationality underpinning this cosmic order. The third topic of

'Trees', was an intermediate stage that focused on the development and diversification of life as a knowledgeable process that became increasingly organic.

In this context, I start with the Kogi notion of the World Tree or Cosmic Pillar as a cosmological framework for the tree-like ecological purpose and development of life. I show how this is instantiated by the origin of trees, their overall functions, place in the landscape, and correspondences with humans. In this context, the Kogi way of understanding and relating to trees, focused on diversification, sustenance and growth, is also a process of 'sowing' and learning 'threads' or literally branches of knowledge. A story about particularly important trees, the Kagksouggi, then indicates an essentially food-like nature of human-tree relations and how the framework of life provided by the World Tree is sustained through nourishing exchanges (Parra Witte 2020). Consequently, many Kogi teachings, responsibilities, and practices are directed towards engendering and maintaining life as human 'seeds', just as the Kagksouggi are primordial trees. To conclude, I return to the life framework of the World Tree and the moral lessons that, in Kogi terms, consolidate environmental care-taking.

The tree-like organisation and growth of life

I had come to the Sierra Nevada de Santa Marta to clarify Kogi ecological knowledge and practices within their wider ontological disposition toward existence. I worked mainly with Mama Shibulata around the Kogi village of Dumingeka, and separately with Mama Luntana and Mama Manuel in Tungeka. All three men performed spiritual divinations or 'consultations' to determine what we should discuss and how. We always held our learning sessions at sacred sites, where spiritual entities guided this process via *aluna*, the universal consciousness or 'thought' and life-force. Mama Shibulata focused more on the cosmological origin and establishment of trees, providing a mythical and explanatory framework for how Mama Luntana and Mama Manuel taught me about the nature of trees and their ecological importance. Both forms of teaching illustrated Kogi dispositions and practices related to trees within their wider role in the Sierra Nevada and the cosmos.

Kagkbʉsánkua: The World Tree

When previously discussing the creation and order of Earth, the Mamas had described Kagkbʉsánkua, the Cosmic Pillar or axis around which everything in the cosmos is fastened. Standing at the 'Heart of the World', the Sierra Nevada, Mama Manuel introduced Kagkbʉsánkua as the first tree, the One, father of all trees. Its extending and multi-layered branches hold the nine cosmic levels around its trunk, which connects the above and the below. The tree's tip reaches the heavenly, solar realms, from where Kagkbʉsánkua is held in place. Its central branches are the cosmos' fifth and middle level, the Earth, whose basic layout is a disc strengthened by an internal cross, and its endings are the world's four corners. Kagkbʉsánkua's roots form the equally cross-shaped cosmic base Shikuákalda (Fig. 4.1), which is reproduced by the frame of the Kogi loom for weaving clothes. Mama Manuel referred to all things in the world as this tree's 'flowers and leaves', because 'everything is attached to it!'. As a tree moreover, Kagkbʉsánkua powers and regulates the global water cycle. After all, Mama Luntana added, trees are key to the circulation of water.

Both Mamas then explained that this tree was 'planted' by the same Four Fathers of creation that built the Cosmos and laid out the Earth: Seizhankua, Aldwáñiku, Siókokwi, and Kunchabitawêya. Accordingly, Kagkbʉsánkua grew upwards to form the nine cosmic levels and to support the world, and subsequently diversified outwards along its branches and twigs into all current life forms. The Four Fathers thereby 'sowed the seeds' for the development of life, said Mama Manuel. While sitting there and picturing all this by the pond at our sacred site, I remembered the all-embracing World Tree or Tree of Life mentioned in other cosmological accounts around the world. I briefly shared this with Juan in Spanish, my Kogi translator at the time, who passed on my query to Mama Manuel in Koggian.[2] He nodded and affirmed that our discussion concerned the same thing; 'that's it, good'.

Eliade (1959, 1961, 44) describes how in many traditions 'the most widely distributed variant of the symbolism of the Centre is the Cosmic Tree, situated in the middle of the Universe, and upholding the [...] worlds as upon one axis'. In some places, like Vedic India or ancient China, the World Tree holds three main cosmic levels, its branches reaching the heavens and the roots descending into the underworld. In other traditions, such as Central and North Asiatic mythologies, the World-Tree holds seven or even nine worlds, as in Kogi cosmology. The tree's Nordic version, called Yggdrasil and described in the Icelandic Prose Edda (Sturluson and Anderson 2015), is moreover aligned with the four cardinal directions, as Kagkbʉsánkua is. Unlike Kagkbʉsánkua however, Yggdrasil has three roots and not four, as the Kogi base-cross Shikuákalda suggests. In various contexts, such as among the Semang of the Malay Peninsula, the trunk of the World Tree is the centre of the 'Cosmic Mountain', whose summit is the World's Centre (Eliade 1961). The Sierra Nevada, considered the 'Heart of the World', is said to have a particular peak that is the 'home' of Kagkbʉsánkua, the Cosmic Pillar.

Different to Rival's (1998, 1, 2, 9) contention that trees are 'natural symbols' used to make concrete the abstract notion of life and eternity, the Mamas speak of Kagkbʉsánkua as a creational and ecological reality known and heeded by the Kogi. Similarly, trees are not metaphors of identity, continuity, social organisation, the body (Rival 1998) or growth (Rival 1993, 649), but embodiments of, and key agents in, an ontological order of life that defines trees, humans and other natural elements in corresponding and interrelated ways.

Threads and branches

One Kogi understanding of the world is as a Fabric of Life (Reichel-Dolmatoff 1978) that binds all elements of creation in a great pattern of 'threads' (*shi*). Their development from a common origin is shown in Kogi bags, which are knitted outwards and upwards from the bottom centre. The rod of the Kogi spindle reproduces Kagkbʉsánkua, the yarn around it are the branches, and the whorl is our middle level of the cosmos. The clothes woven on the cross-shaped Kogi loom are the earth that covers the structure of this middle level. By this twisting, knitting and weaving of threads, Kogi men and women remember and follow the fibres through which the Great Mother of creation 'thought' the world, subsequently planted by the Four Fathers.

Kagkbʉsánkua's many branches and twigs then, are the threads of the Fabric of Life. They ramify in complex derivations that form an overall cosmological structure. Its common origin are the roots and the trunk. Each branch or thread is a unique reproduction, aspect, or variation of the whole tree or fabric in a pattern of expanding self-similarity. Halbmayer (2020, 16) gets close to this logic when arguing that the 'all-embracing ontological continuity' exhibited by the peoples of the Sierra Nevada is 'homological'.

Reichel-Dolmatoff (1985a, 1985b) tells us that the Kogi also speak of the genesis of humanity and all things as the growth of a squash plant, whose stalk is located in a place called Cherúa. 'Branching out more and more' in all directions across the territory and according to the Law of the Mother, the plant's branches and fruits populated the Sierra Nevada's valleys. 'Some died, but in their place new ones were born. Others dried up, but others grew strong and healthy'. They gradually got 'farther and farther away from their origin, but always remaining united with the Mother 'because she was the first trunk, from which all had come'. Since She created humans, the animals, plants, 'in sum, the entire universe, everything that forms part of it [...] has its place in this immense genealogical tree' (Reichel-Dolmatoff 1985a, 155, 156; 1985b, 86).

Kogi lineages are part of the Sierra Nevada's diversification of branches or threads. Each lineage is associated to specific cosmological principles and natural elements that constitute complex creational collectives or 'families'. These descent groups are derived from common ancestors called the Spiritual Fathers and Mothers (*Kalguasha*), a kind of 'cosmocratic god-persons' (Sahlins 2014, 286). They give rise to, and sustain, 'different things that occur together' (Halbmayer 2020, 17) including humans, animal and plant species, landmarks, celestial bodies, elements like fire or wind, or substances like blood or shells. The corresponding objects, stories, rituals, knowledge specialisations, and social institutions of each Kogi lineage are aligned with these creational associations and maintain the ecological well-being of their families of natural elements.

Shi (thread) is also the stem of *shibɨlama*. It is the knowledge in and of things that, as a Kogi friend named Alberto once broke down to me, constitute the hot threads of truth and life that come from *aluna*. They are also 'paths' by which the Mamas referred to our topics of water, earth, trees and food. The Kogi additionally use *shibɨlama* to designate what may be termed 'culture', that is, a people's knowledgeable way of being, thread or branch in the wider Fabric or Tree. *Kaggaba chi shibɨlama* is 'the knowledge of the Kogi', and *Pldañsé chi shibɨlama* is 'the knowledge of Europeans'. As Mama Manuel said:

In the beginning, the Mother gave each group of people their own *shibɥlama*, customs, language, dress, and place to live.

For this reason, all peoples are 'children of the same Mother', ramifications of the same trunk with shared roots. *Shibɥlama* is ultimately one and the same, merely taking different forms in different places and peoples, but originating in the Sierra Nevada at specific sacred sites. Eliade (1961, 43) notes that in many traditions 'the creation of man [sic], a replica of the cosmogony, took place similarly from [...] the Centre of the World'.

Among the Kogi, human cultural and geographical variation is one manifestation of cosmological structure and diversification, which manifests as an ecological web of relations driven by knowledge. The Kogi thus seem to merge the notion of 'World Tree' with its related variations as 'Tree of Life' (Giovino 2007) and 'Tree of Knowledge' (Fernandez 1998) into one integrated understanding. Furthermore, 'planting' Kagkbɥsánkua and 'sowing' *shibɥlama* are aspects of a creational, integrated correlation between growth, order, and diversity. While through trees the Amazonian Huaorani People understand, express, and relate to general principles of growth and well-being that combine the natural and the social, the Kogi embed these principles in an elaborate metaphysical system. Here, trees embody life patterns as reproductions of Kagkbɥsánkua, the primordial prototype.

Morality, knowledge and ecology

Taking the discussion further, Mama Manuel emphasised that the reason Kagkbɥsánkua was 'planted' by the Four Fathers of creation is to support the world. Similarly, *shibɥlama* was 'sown' for humans to learn how to care for the world. Expanding the archetypal notion of a 'tree of enlightenment or re-invigoration' that gives 'special power and insight' and is related to fertility (Fernandez 1998, 83), the Mama added that 'we are all in charge of this' care-taking work. After all, everyone has 'the same aluna' or spirit/thought, and each human group performs this shared duty according to their own thread or branch and in their own place, together sustaining the whole Tree. Since the Kogi consider themselves to be 'elder brothers' of humankind (Reichel-Dolmatoff 1985a, 211) and the Mother's 'first children', they are in charge of the trunk, the Sierra Nevada, on which Kagkbɇsánkua's stability depends. With this greater responsibility also comes 'deeper' knowledge, a *shibɇlama* closer to the tree's roots and hence the fabric's origin. 'Why is your knowledge deeper?', I asked Alejo, a young Spanish-speaking Kogi who helped me understand the Mamas' words in Koggian. 'Because the Mother gave us the Sierra to look after, and the Sierra is the Heart of the World that contains everything'. However, Mama Manuel reminded us, 'many people are not fulfilling their duty to care for things anymore'.

This 'planting' and 'sowing', he continued, was for everyone 'to live well'. *Shibɇldama* is there 'for us to value this existence', 'think and feel like humans' and behave correctly. This knowledgeable, ecological morality then, promotes the growth of trees, plants, and everything else. Reichel-Dolmatoff (1978, 13) was similarly told that weaving a finished textile is an act of discipline, good thinking, and a well-led life. The Fabric of Life is a 'web of knowledge made of [good] thoughts, [...] life's wisdom envelops us like a cloth'. The correlation between the Fabric of Life and the Tree of Life that I discovered, means that the

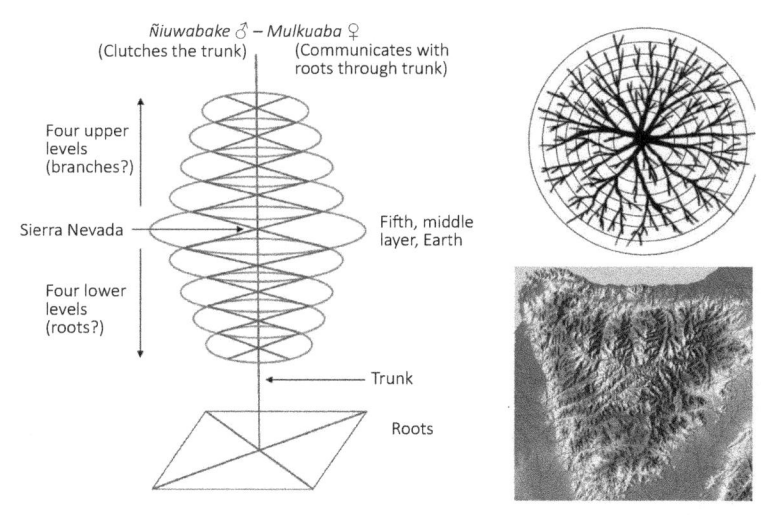

Fig. 4.1 The structure of the cosmos as a World-Tree (Kagkbɇsánkua) that also ramifies from the centre of the Sierra Nevada. River valleys and their corresponding human lineages are seen as branches.
After Parra Witte (2018). Map: Wikimedia Commons.

moral aspect of weaving is intertwined with the ecological sustenance of tree-like development, guided by the same knowledgeable process.

The creation and role of trees

Having laid out this creational tree-framework, a few weeks later the three Mamas moved their narrative to the establishment of trees themselves. This happened after the Earth itself was organised physically from its previous conception in 'thought' or *aluna*, around the time of the world's 'dawn' into the sun's light. Mama Shibulata stated that the land was still empty and barren, its surface being devoid of growth. 'Something was missing', something of vital importance for the future. This was trees and vegetation. This planting, Mama Luntana and Mama Manuel asserted, covered what had until then been a 'bald earth', and started a cycle of growth that 'healed the Mother' by 'removing' the Earth's initial rough conditions, such as fire, landslides, earthquakes, or erupting mountains.

Birth at sacred sites

Talking under trees and amidst bushes at our sacred sites, I was taught how, as all natural elements, trees were distributed in an organised fashion from the top of the Sierra Nevada downwards and outwards, and therefore from centre of the World Tree along its expanding structure.

The spots where things came into existence are the sacred sites, which form a complex, concentric network or grid covering the Sierra Nevada. Being places where life originated, each of which sustains specific natural elements, the Kogi define sacred sites as territorial 'fathers' and 'mothers' (*hate* and *haba*) of animal and plant species, marriage, sicknesses, traditions, blood, or anything else. Sacred sites are therefore nodes of 'threads' in the Fabric of Life, and serve as ritual and gathering places for the Kogi. Most sites I visited are marked by trees or bushes in special locations, having unique characteristics, or contrasting to an otherwise tree-less area. They combine with (clusters of) stones and rocks of different sizes, shapes, and positions. If a place has both special stones and trees,[3] it most definitely is a sacred site. When talking, teaching, divining, or ritually reciprocating nature there, the Mamas usually sit at the most central spot on the site on a rock and/or by a tree. Being nodes of spiritual energy and knowledge that power and guide nature, it is 'forbidden to plant trees' at sacred sites.

Following how Father Seizhankua laid out the earth and its different soils, the diversification of trees equally spread from a class of primary sacred sites or creational nodes called *eisuamas* that 'organise' life. This also makes them 'places of government' where the Mamas learn, the Kogi guide their traditions, and essential care-taking ceremonies and practices take place. The creational families introduced earlier, which include Kogi lineages as 'branches' of the World Tree, thus have their ecological and spiritual centres at corresponding *eisuamas*.

The principal origin of trees lies in a female place called Dumena, one of the Great Mother's first 'daughters', who also engendered lakes. There, the Father and Mother of Trees, Kalakshé and Kalawia, were 'born' and prepared the 'sowing of trees'. These Parents had previously conceived trees in the realm of *nuhuakalda*, a sort of otherworld inside mountains and accessible through caves. *Nuhuakalda* is the spiritual storehouse and source of sustenance for all that exists. To see which places and groups of 'ancestral' Kogi needed or wanted trees for the future, in a hierarchical succession according to importance Kalakshé and Kalawia then went from one *eisuama* to the next to 'offer their trees'. At the time, those trees 'were like persons' in spiritual form, specifically 'women'.

Offering, baptising, and planting

Sitting on his usual rock at our sacred site, Mama Shibulata described in great detail the events in each *eisuama*. He vividly recreated the response of those places that 'did not like these trees/persons', either openly rejecting them or 'buying' only a few. *Eisuamas* that partially accepted trees, bought or 'received' a somewhat greater number. Foreseeing that if this rejection or only partial reception of trees continued the Earth would remain mostly bare, a Spiritual Father called Mulkueke instructed the remaining *eisuamas* to receive many trees. The inhabitants of those places consequently realised the future significance of trees for humans and the world. 'If not for that', Alejo added, 'there would not be any trees in the world today'. When walking the Sierra Nevada, 'one may wonder why certain areas of the Sierra have less trees than others, or drier conditions'. Slopes and valleys around *eisuamas* that received either none or a few trees are scarcely forested. Where trees were partly accepted, wooded areas cover only some mountain tops and river sheds. *Eisuamas* that acquired many trees are surrounded by denser tree-cover.

The Mamas in Tungeka clarified that the main reason for not receiving trees, was that back then people living at *eisuamas* were powerful enough to simply 'make appear whatever they needed, such as wood and fire for cooking'. Knowing that trees could be spiritually obtained from the realm of *nuhuakalda*, Mama Luntana explained, some *eisuamas* ignored the need for naturally occurring trees. They failed to see that in the future people might not have those powers anymore and could depend on trees for their materials. This is why Father Kalakshé and Mother Kalawia were offering their 'children'. Initially diverging or conflicting forces became a lesson for future generations and allowed a functional vegetation cover. 'Buying' trees moreover, expresses the Kogi understanding that ecological exchanges or dependencies are akin to economic transactions (Ereira and Attala 2021), in the sense that they are pragmatic obligations to 'pay' or give back what is received from nature to maintain balance.

Once offered, Mama Shibulata carried on another day, the next step was to 'baptise' trees. This term was borrowed from Christian missionaries in previous centuries to express in Spanish the Kogi practice of initiating something or someone into physical existence from its origin in *aluna*. Being female spiritual persons, trees had to be turned into organic beings, just like lakes and soils are also materialised spiritual girls that underwent a 'maturation into full-grown women'. This initiation, and hence fertilisation of trees, was performed by the male character Dibʉndshizha. To be transformed, the female trees underwent a learning process of 'nine stages' high above in the celestial realms where the Spiritual Fathers and Mothers reside.

Trees were then 'sown' and 'planted' for the first time in a sacred site near Hukumeizhi, an *eisuama* that received many trees. The place is called Kaxsʉma, who is a 'mother' that provided the 'seeds' for trees to grow. Mama Manuel and Mama Luntana told me separately that she spiritually obtained these seeds from the otherworld *nuhuakalda*, and

Table 4.1 The eisuamas' different reactions to being offered trees, in approximate order that they appear in the storyline

Eisuama	Reaction	Notes
Sʉxdzibake	Did not like/want trees	
Magutama	A little bit	
Alduaka	Nothing, not needed	
Zâldaka	Not needed	
Hubiskuo	A little bit, not more	
Sekandzhi	Did not like, just a couple	
Seizhua	More or less	
Hukumeizhi	Accepted many trees	Here Mulkueke instructed ezwamas to receive trees. The place where trees were later 'baptised'.
Nabbʉguizhi	Bought many trees	
Nuabaka	Also accepted trees	Previous three eisuamas agreed to buy many trees
Kuamaka	Just a bit	
Giumaldaka	Not much	
Abbleizhi	A little bit	
Indshizhaka	A little bit	
Súguldu	A lot	
Mamaldwúa	Bought many trees	Agreed to do so together with Hukumeizhi

After Parra Witte (2018).

then 'extracted the seeds from within herself, [...] like the seeds that human women produce for birthing children'. Finally, she asked all tree-persons: 'my children, what will you be good for?', each of whom replied 'I will do such and such'.

Distribution

Having been baptised, taught, and materialised, physical trees could then be distributed around the Sierra Nevada according to the eisuamas' preferences. After Mother Kaxsʉma gave the seeds to another male character, seemingly Seizhadzhíñmako, he started planting them from the Sierra's centre, where Kagkbʉsánkua's trunk is, through its valleys and 'until the end of the world'. Following Mama Manuel and Mama Luntana, all trees in the world communicate with this World Tree, being 'like his children, spread all around him'. In this vast, complex ramification of tree species, habitats, and sacred sites converging around Kagkbʉsánkua, he is both the model for all trees and the cosmic order according to which they were created.

As on the Sierra Nevada, trees in other regions were offered to sacred sites, being also variously rejected or accepted, this giving rise to the forested, semi-forested, or bare areas of today. 'Perhaps in the Sahara none were wanted!', Mama Manuel intervened. Seizhadzhíṅmako apparently not only sowed trees but also bushes and plants across different landscapes around the world. He then 'named' the different tree species, assigning them future habitats and functions at sacred sites that are now their 'fathers' and 'mothers'. At those places, Seizhadzhíṅmako also deposited corresponding knowledge or *shibʉlama*, so that like Kagkbʉsánkua, the establishment of trees and plants was also a set of 'lessons' or 'teachings' that tell humans how to know, use, and behave toward these entities.

Initially however, the world was covered with vast forests, something unsuitable for the development of life. Trees were so tall that they reached the clouds, which could have inappropriately allowed humans to climb up to the sky. The atmosphere in these extensive, dense forests under a thick canopy was a dark and inhospitable one where dangerous and giant animals thrived, as well as man-eating spirit-beings. The number of trees inhibited the growth and flourishing of food plants and crops. To create better living conditions, Father Mandáouldo reduced these forests, resulting in the balanced distribution described earlier. 'So, the story teaches that cutting trees and having too many are unfeasible extremes of a balanced tree-cover?', I checked. The Mamas confirmed this: Mandáouldo instructed humans to care for and keep renewing the tree-cover *as he left it*.

Protecting life-forces, linking realms

Following a quick divination to find out what we should talk about that day, Mama Luntana and Mama Manuel told me that, 'together with water and earth, [trees] sustain life'. After they told me how trees produce fresh air, regulate the water cycle, provide food and homes for animals, serve as wood, or have medicinal properties, I felt compelled to ask: 'why do trees exist in the first place? Why did the Mother give them to the Earth?'.

Mama Luntana responded that the tree-cover on hills and mountains corresponds to the straw used to thatch Kogi houses, including the men's ceremonial world-house, the *nuhué*. Just like the straw protects the interior of a house, the tree-cover protects what is inside hills or mountains, the otherworld *nuhuakalda*, and helps contain its stored life-forces. This dark realm corresponds to the space inside the *nuhué*, whose architecture reproduces the structure of the cosmos. Furthermore, 'all knowledge' (*shibᵾlama*) comes from *nuhuakalda*, it is 'where the life and wisdom that the Mother gives us comes from, her teachings'. Since especially certain mountains are full of this knowledge, this is why 'mist, rain, and air form on top of them'. So, 'if we cut these trees down', or worse, 'if we mine or chop off the hill itself', its *shibᵾlama* 'spills out or evaporates'. Those hills or mountains moreover, 'are the nuhués' of spiritual people, *aluna kággaba*, so that felling the trees on top is like destroying the roof of their homes in *nuhuakalda*. 'Feeling unwell', they wonder: 'why are you removing our thatching?'.

Conveying the Mama's words, Alejo then explained that another main purpose of trees is to connect the mountains' interior, *nuhuakalda*, with physical nature outside. While the roots of trees are a link into Mother Earth's body below, their branches and leaves communicate with the air, the sun, and the sky above. In this way, trees emulate Kagkbᵾsánkua, the World Tree, who binds the nine cosmic layers from its deep and dark roots, the base-cross Shikuákalda, up to the celestial and light-filled realms of Ñiuwabake and Mulkuaba (Fig. 4.1). Trees 'are like antennas that transmit everything' in both ways. 'Through the air we breathe' then, significantly produced by trees, 'we learn from them and receive the shibᵾlama that they transmit out from nuhuakalda', which guides 'our well-being'. This is partly why the Mamas are traditionally trained in caves, being entrances to the spiritual otherworld.

Trees and humans

Following Rival (1998, 9, 11), the identity-giving, 'symbolic' analogy between trees and human bodies is a recurrent theme across peoples, and often includes non-material aspects of the human person like self, spirit, or personality. Resonating with this, Niño Vargas (2020, 45, 46, 49) argues that among Chibcha groups, trees and plants are analogous to humans in shape and process, while pertaining to a divine vegetal past that serves as a metaphorical exemplar for present, true humanity. Rather than ontologically emerging *from* flora (Niño Vargas 2020, 51), as the Kuna of Panama and Colombia believe (Chapin 1989, 152), in Kogi terms trees and humans developed from the Spiritual Parents, who are prototypes of all natural elements. This is why, like trees and other natural elements, the Kogi were first spiritual persons (*aluna kággaba*) who were then materialised. Most importantly, trees and humans and are both part of a wider territorial order that defines their relationship.

Ontological correspondences

Elaborating on these parallels, Mama Luntana explained that during the Earth's materialisation, hard elements such as rock and sand came first, then came the soil, and finally vegetation. Likewise, humans were first made with bones (our rock), at which point the Four Fathers saw that veins were needed (blood = oil), yet veins could not be held without flesh (= soil), which could not be contained without skin (= surface), and this required hair (= trees) to protect the skin. Just as a hitherto 'bald' Earth was covered with 'its hair', initially hairless humans underwent capillary coverage when turning from spiritual persons to physical beings. While trees were planted, our hair grew. More specifically, Mama Manuel said that because human armpits are like valleys, they are 'wooded' and produce 'rivers' (sweat). In the case of animals, fur, scales, shell, or just skin also correspond to the Earth's different surfaces. Since the Earth itself is the body of the Great Mother moreover, everything on Her surface can be seen as Her 'hair'. Unnecessarily cutting or plucking out one's hair, therefore, is like unduly felling trees and forests without permission or spiritual preparation.

Grabbing his ponytail, Alejo added that 'we humans have hair' because it is equally a channel of communication between our internal soul, 'our aluna', and the external environment. Just as *nuhuakalda* sustains the Sierra-body and holds its consciousness, the *aluna* of humans is the source of their *shibʉlama* (knowledge), vitality, and thought. Because 'we are like the hills and mountains', Mama Manuel intervened, the *shibʉlama* stored inside us is 'what we think with'. For the Kogi, humans organically resemble the Sierra in body and spirit, whose relationship is mediated by hair and trees respectively.

Apart from hair being like trees, Mama Luntana and Mama Manuel continued, 'trees used to be persons like us'. In this light, the orderly distribution of trees is equivalent to that of human groups. This depends on how their ancestors treated trees according to the

Fig. 4.2 Seizhua, Mama Shibulata's *eisuama* (top left), partly accepted trees. Before the *nuhué*, stands a sacred coca-leaf tree. Mama Salé, who grows facial hair, hails from the forested Hukumeizhi valley (top right). Mama Bernardo (bottom right) is from the *eisuama* of Kuamaka (bottom left), which accordingly received no trees (houses are on the small plateau).
Photos: Falk Parra Witte and Bernabé Zarabata.

story above. In the Kogi case, different lineages across the mountains have more or less hair according to which *eisuama* they belong to, and thus how forested their valleys are. Personally, I saw few Kogi men with facial hair, which in this logic is because I spent much time in an *eisuama* valley system that received few trees. While 'you will not find anyone with facial hair around these parts', Alejo said, 'some men from the forested Hukumeizhi valley, which accepted trees, grow beards, moustaches, or goatees'. The Hukumeizhi lineage is consequently in charge of trees and spiritually communicates with their Spiritual Parents, Father Kalakshé and Mother Kalawia.

Another equivalence between trees and humans further exemplifies how Kogi collective identity is embedded in cosmological organisation and geographical conditions. On the Sierra Nevada, both Mamas shared with me, there are 'Kogi trees' and 'non-Kogi trees'. Just as the Kogi consider themselves to be 'elder brothers' of humankind (Reichel-Dolmatoff 1985a, 211), Kogi trees, the 'first to be born', are hence located in higher parts of the Sierra (above approximately 1000m), and are usually smaller. Like most humans, non-Kogi trees are like the 'younger brothers' of trees, being found at lower levels and normally larger. Since elder brothers and younger brothers 'should not mix and live in each other's territories', Mama Manuel asserted, trees should do the same. For example, 'planting non-Kogi trees higher up and Kogi trees further down can kill them'. The Mamas then discussed the names and traits of these tree types, which they easily identify when walking the mountains.

'Do other peoples also have corresponding trees, say German Trees?', I felt compelled to ask. They replied that this is not the case, and confirmed the same, basic duality. 'Kogi-trees' refers to 'indigenous trees' in general, being distinguished from non-Indigenous people and their associated trees. Still, all trees have the same origin, the same parents, Kalakshé and Kalawia. As elder brothers, however, only the Kogi were provided with the seeds of trees, so that 'we could plant them around the world'. 'Other indigenous peoples are in charge of caring for trees, but we also planted them'. One night by the library at the Kogi school in Dumingeka where I worked for a year, a Kogi teacher named Juan Manuel affirmed that the Kogi are literally human seeds. 'This is why we are elder brothers', and for things to grow, 'seeds cannot be replaced'.

Table 4.2 Examples of Kogi and non-Kogi trees. Their botanical names were not identified.

Kogi trees	Non-Kogi trees
Gaxálda	Ikaldá
Geizhá	Mikuédzi
Kandzhí	Mitábi
Kinakina	Moúxlda
Kokuizhá	Mis̷ɨ
Nabí	Shiná
Saxíndla	Taduá
Taízhi	
Tamshé	
Tuldá	

After Parra Witte (2018).

Fig. 4.3 Tall 'non-Kogi trees' at about 200m above sea level (top) stand in contrast to smaller 'Kogi trees' surrounding a traditional homestead at an altitude of approx. 2.400m (bottom). The imposing tree on the top left is a Kagksouggi tree on a sacred site, where a group of Mama are divining.
Photos: Falk Parra Witte.

The Kogi and the Kagksouggi

In her book *Finding the Mother Tree: Discovering the Wisdom of the Forest*, forest ecologist Suzanne Simard (2021) argues that, as key agents in the circle of life, trees are communal, cooperative creatures that learn, perceive, and communicate. They do this for instance via complex underground networks. In particular, Simard has identified specific large trees termed 'mother trees' that act as central hubs for these networks, and interconnect and sustain the forest around them. Talking about the distribution of trees, Mama Shibulata specified something very similar. Mandáouldo, who had organised the tree-cover, left behind certain types called *kagksouggi*. Located at key places in the interconnected and *living* network of sacred sites, these particularly sentient, old, and important trees are considerably fewer than the rest. The Kogi can easily identify these special trees, and Alejo explained that even if out of eyesight, people can sense a *kagksouggi* in a nearby grove. Such trees stand guard over the trees and woods around them, being like centres of sustenance. In contrast to other trees, the *kagksouggi* remain 'people', having retained their personhood up until today. *Kagksouggi* are usually accompanied by different kinds of large stones, who are conscious, petrified 'ancestors'. For these reasons, felling *kagksouggi* is prohibited, causes harm to their surrounding dependent trees, and creates negative consequences for the human perpetrator.

My sessions with Mama Shibulata normally took place at a secluded sacred site amid lush vegetation near his home, where his family members sometimes dropped by while we talked. One time, the Mama unexpectedly chose a more exposed sacred site where other Kogi and non-Kogi people walk past. Despite these noises and movements, he provided a vivid background to the *kagksouggi*'s personhood, animatedly living and reproducing the story's sounds, dialogues and gestures. The Mama's gaze was somehow unfixed, as if mentally visiting another place and time:

> After the world dawned and trees were baptised, Father Kalakshé and Mother Kalawia emerged from the realm of nuhuakalda unto this physical existence. They gave birth to a community of people called Kagksouggi in a place called Tashízhua.

I enquired:

> Did they look like us?
>
> Yes, they did. These equally 'indigenous' people lived contemporaneously with the first Kogi but in separate societies, when humans 'did not yet use trees for building, cooking, and other tasks'.

Although the Kagksouggi People were authoritative, knowledgeable, and powerful, they did not use these faculties for the good. Their leaders tended to mistreat the Kogi in retaliation for the wrong that humans had done to the spiritual trees-to-be by regularly rejecting them when offered to the *eisuamas*. Seeing them as 'unnecessary', 'hairy', and 'ugly', the Kogi had insulted them. Alejo compared this to when current Kogi headmen demand food from villagers who disobey or perform inappropriately. However, the Kagksouggi's revengeful behaviour was also mean and further intensified the problem.

Kagksouggi headmen deceived and 'ate' Kogi men by intercepting them on their way to the *eisuamas* to 'confess' with the Mamas. Confession is a term borrowed from Christian missionaries to refer to a Kogi practice called *aluna zhiguashi*. The Kogi mentally discharge their thoughts, actions, and feelings at a sacred site to cleanse their energies, harmonise them with the environment, and nourish natural elements. The Kagksouggi offered the Kogi the apparent opportunity to do this, but in reality, this was a cunning trick to 'open' or 'untie' the distracted Kogi's heads at the top, 'where all our body is tied-up like a bag [to keep] the contents inside', and extract everything. Since the Kogi were food/nourishment for them, the Kagksouggi used their powers to turn these contents into crops (e.g., plantain, cassava, potatoes). Knowing the lives of these Kogi individuals by having heard them confess, the Kagksouggi headmen then put on the remaining, empty skin or shell to impersonate the man and deceive his wife and fellow villagers. When people were asleep, the Kagksouggi left the shell and filled it with ash as an apparently resting body.

Mama Shibulata outlined how Kogi Mamas and headmen defended themselves from the Kagksouggi, 'fighting with words' to determine who was 'better', get rid of the other, and 'win', just as Mamas may nowadays 'compete for status or influence'. The struggle involved powers and abilities like disappearing, clairvoyance, or golden staffs that transformed things into fire or stone. Indignant at the Kagksouggi's actions, the Kogi sent wise and powerful characters to perform the same confessions and extractions on the Kagksouggi, but each one was defeated. It was the determined Sintana who finally managed to evade the Kagksouggi's deception, capture seven individuals to convert them into food, and trick the Kagksouggi community into thinking he was the defeated one. Having previously visited the Mamas at the *eisuamas* in cycles of seven days to confess and receive their spiritual protection against being turned into food, Sintana gradually diminished the Kagksouggi people by taking away their powers and ritual objects. 'Bad people never end up winning', Alejo concluded.

In line with the overall birth of trees, the Kagksouggi then became physical trees. Mama Shibulata explained that because the Kagksouggi abused and harmed the Kogi, humans nowadays (have the right to) cut trees and use their wood. The Mamas in Tungeka added that this should be achieved in measured ways; people must respect the number and place of trees, ask for 'permission' to fell them and/or acquire their resources, and in return perform spiritual 'payments'[4] to the Parents of Trees. According to Mama Shibulata however, humans are unknowingly repeating the Kagksouggi's behaviour, currently abusing trees and felling them in great numbers without respect. Instead of learning from these past events, we are caught in a vicious cycle. 'If we continue doing this', Mama Manuel and Mama Luntana warned, 'one day trees will become people again, and return all our harm'.

A shared origin, substance and purpose

The Kagksouggi story seems to exhibit elements of the Amazonian type of animism known as 'perspectivism' (Viveiros de Castro 1998; 2004), given how trees and humans are both persons that, from their own point of view, simultaneously perceive the other as food. More than that, this relationship is based on a reciprocal exchange of equivalent substances in which each side has to respect the other.

I asked Mama Shibulata if the reason Kogi confessions nourish is because the Kagksouggi turned people into food. 'Indeed', he replied, 'that is why when we confess spiritually with words, we send a payment to the (celestial) home of the Father and Mother of Trees', communicating with them. 'In aluna, they see this as food', which ensures that these Parents always sustain physical trees. 'Is that why the Kagksouggi could transform people into food? We are food'. I confirmed. Yes, 'food, that's true', came Alejo's reply. Following Mama Luntana and Mama Manuel, the Kogi pay trees, both in general and specifically the Kagksouggi at sacred sites, for them to keep growing, not dry out, and remain useful to humans. By thus nourishing them, the Kogi continue to be food for trees, who in the spiritual realm are still persons. Because the Kagksouggi were equally turned into food, trees now nourish humans and provide them with materials. In this vein, to pay for the teachings I received about trees, I brought gifts of food to the Mamas, who in turn paid the Parents of Trees for this knowledge.

As mentioned before moreover, the Kogi see themselves as 'seeds', that is, 'elder brothers' of humanity who have the responsibility to care for all things, and who also planted trees. Their cultural 'thread' of knowledge is closer to the cosmic trunk of the World Tree Kagkbʉsánkua and the roots of growth. For this reason, at sacred sites 'we confess and pay to all things, because that is their food. We Kogi are in charge of planting and caring for everything, so [that all beings] can live well'. *Kagksouggi* trees are equally considered 'seeds' that were sown 'from the origin' at sacred sites on the Sierra Nevada, the 'Heart of the World' from where Kagkbʉsánkua's branches ramified (or of the cosmic squash plant). Accordingly, *kagksouggi* generally indicates 'something ancient', something that grows 'from the source'. I explained that beyond the functions of all other trees, *kagksouggi* are also centres of support at sacred sites for their surrounding environment. As a result, *kagksouggi* trees and Kogi humans are both 'people' that share not only a spiritual essence, but also a common origin and purpose.

In this context, what ends up defining personhood is the enactment of ecological roles or functions that are learned and implemented to regulate primordial life-forces. The story suggests that conflict should be replaced by a conscious, mutual exchange that renders the shared distribution and ontological equivalences of trees and humans a matter of ecological sustenance. More than trees expressing the cyclical and reproductive character of human society, as among the Amazonian Huaorani (Rival 1993, 636), Mama Luntana indicated that the *shibʉlama* or knowledge of trees instantiates the structured, purposeful growth and re-growth of all things. Nourishing is equal to the 'planting' and 'sowing' of this knowledge, which the Mamas learn by communicating with *aluna*, the thought of nature and also the medium of nourishment. It is understandable then, that on special ceremonial occasions the Mamas wear sacred masks or wooden faces (*kaldwaká*) that embody the organic growth of trees (*kaldyi*) and personify the Spiritual Parents in order to reproduce the creational forces of these characters. All in all, this knowledgeable ecological sustenance maintains the life-giving order of branches, leaves and flowers of the World Tree itself. 'If we do not understand and respect the place and purpose of all things', Mama Manuel concluded, they would not exist, and 'trees and plants would not grow'.

Conclusion: Watering Kagkbʉsánkua

It was noon and very hot. Sweating considerably, Alejo, Mama Manuel and I were sitting between the bushes on uncomfortable stones with no escape from the scorching sun. Nevertheless, the Mama's words were very inspiring. Mama Manuel rounded up the lessons on trees and plants by returning to the initial life-framework of the World Tree, Kagkbʉsánkua.

Learning *shibʉlama*, he began, is like mentally trying to cut down a very large tree with an axe. Since this knowledge is about the origins and ways of being of everything, progress is slow and 'one advances hack by hack', so that at times the person 'may become tired or disheartened'. But 'you have to be strong Falk, always continuing in spite of the hardness of the tree and the great amount of work'. However, since this Tree of Knowledge was planted by the Four Fathers, 'it will never really be cut down, since no one will ever manage to know all shibʉlama'. People can only access parts of this tree, including the Mamas and their apprentices, who are 'like axes'. Instead of trying to bring down a trunk that is too thick anyway, the idea is to keep learning, listening, and striving to emulate the Spiritual Fathers by continually working at their tree. 'These sessions are a start', and 'with the wood that you cut from this giant tree, you should build a house for the shibʉlama that you are learning'. In this mental home, the knowledge can be kept and nurtured, just like the *nuhué* holds Kogi men's thoughts and discussions, or how *nuhuakalda* holds the Mother's life-forces and knowledge.

'The axe', Mama Manuel elaborated, is the Kogi way of learning and doing, which involves patience, effort, and perseverance. Here, the process of cutting knowledge is more important than the goal of felling the tree. By contrast, the younger brothers behave 'like chainsaws', wanting to cut down trees and dominate everything quickly and easily, which leaves little space for lessons and experience. Chainsaw behaviour is 'disrespectful' and mindless, looting and harming here and there without stopping to value and recognise the knowledge inherent in things. What is cut is later simply discarded. One Kogi designation for non-Indigenous people, *zhaldzhi,* accordingly means something like 'the ones that eat everything'.

By thus damaging the Mother's body, the Earth, we also harm the World Tree, Mama Manuel continued. Because Kagkbɵsánkua perceives everything that people do, he 'sees and feels how we peel his branches, cut his twigs, pull out his flowers, and dry out his leaves'. Altering the balanced life-framework of the World Tree causes conscious reactions in nature like earthquakes, changed weather patterns, or species reduction. If Kagkbɵsánkua supports the world we live on and is 'the only one', then 'why do we harm him?'. Yet humans not only destabilise Kagkbɵsánkua through activities like mining, damming, polluting, displacing, and logging. He is also harmed spiritually by people's thoughts and attitudes when they do not value, respect, and learn. This manifests as social disorder, violence, fighting, bad words, and negative feelings. Being also a Tree of Knowledge, what we do against Kagkbɵsánkua is un-thinking, un-nourishing, and hence un-knowledgeable behaviour. All of this disregards the threads of *shibɵlama,* making people ignore the order of life embodied and sustained by the Tree's trunk and branches. These conditions affect Kagkbɵsánkua 'as they would affect any person', making him wonder: 'what's wrong with people?'.

People's attitude of continually 'chopping' at this World-Tree (*Kagkbɵsánkua mengwi*), is also what makes them fell more and more trees (*kaldyi mengwi*). 'They are persons that feel what we do to them', something that they could do to us in the future. 'If trees give us wisdom', Mama Luntana added, 'why do we cut them down so much?'. In addition, Mama Shibulata stated, the Mamas see that '[plants] are vanishing, spiritually going back to their mother'.

Despite all this, Mama Manuel ended his moral narrative on a positive note. Valuing, respecting, learning, and caring for things by contrast, is 'watering Kagkbɨsánkua'. Just as trees and plants need to be watered and cared for if they are to grow, so do our positive actions, emotions, and thoughts create the conditions for Kagkbɨsánkua 'to flourish and blossom', allowing 'the world to heal'. By thinking and acting well based on *shibɨlama*, 'we plant trees in aluna', and thus also sustain the spiritual realm. In this manner, 'things can get better'.

In sum, trees and humans are part of a complex arrangement of mutually sustaining spiritual forces, biological cycles, landmarks, and natural elements. It is all structured and vitalised by the World Tree, whose trunk, branches, leaves, and flowers are the cosmic framework and model for life. By ending the lessons with the moral quality of this framework, Mama Manuel also integrated the 'threads' that run through this chapter: namely that tree-like diversification, nourishment, and growth are a thoughtful, knowledgeable process that guides ecological relationality. This integration is synthesised by the Kogi equivalence between primary trees (*kagksouggi*) and primary humans (*kággaba*), who are cosmic seeds carrying teachings that engender and regulate nature.

References

Chapin, M., 1989. *Pab Igala. Historiad de la tradición cuna*. Quito: Abya Yala.

Eliade, M., 1959. *The Sacred and the Profane: The Nature of Religion*. San Diego, CA: Harcourt Brace Jovanovich.

Eliade, M., 1961. *Images and Symbols*. London: Sheed and Ward.

Ereira, A. and Attala, L., 2021. 'Zhigoneshi: A culture of connection', *Ecocene: Cappadocia Journal of Environmental Humanities*, 2(1), 7–22.

Fernandez, J. W., 1998. 'Trees of knowledge of self and other in culture: On models for the moral imagination', in L. Rival (ed.), *The Social Life of Trees: Anthropological Perspectives on Tree Symbolism*. Abingdon and New York, NY: Routledge, pp. 39–55.

Giovino, M., 2007. *The Assyrian Sacred Tree: A History of Interpretations*. Göttingen: Vandenhoeck and Ruprecht.

Halbmayer, E., 2020. 'Introduction: Toward an anthropological understanding of the area between the Andes, Mesoamerica, and the Amazon', in E. Halbmayer (ed.), *Amerindian Socio-Cosmologies between the Andes, Amazonia and Mesoamerica: Toward an*

Anthropological Understanding of the Isthmo–Colombian Area. New York, NY: Routledge, pp. 3–34.

Niño Vargas, J. C., 2020. 'An Amerindian humanisn: order and transformation in Chibchan universes', in E. Halbmayer (ed.), *Amerindian Socio-Cosmologies between the Andes, Amazonia and Mesoamerica: Toward an Anthropological Understanding of the Isthmo–Colombian Area*. New York, NY: Routledge, pp. 37–60.

Parra Witte, F. X., 2018. 'Living the Law of Origin: The Cosmological, Ontological, Epistemological, and Ecological Framework of Kogi Environmental Politics'. (Unpublished PhD thesis, University of Cambridge).

Parra Witte, F. X., 2020. 'The Structure that Sustains Life: Nourishment and Exchange Among the Kogi', *Tabula Rasa* 36, 101–29, doi: 10.25058/20112742.n36.04.

Reichel-Dolmatoff, G., 1978. 'The loom of life: A Kogi principle of integration', *Journal of Latin American Lore* 4(1), 5–27.

Reichel-Dolmatoff, G., 1985a. [1950] *Los Kogi* I. Bogotá: Procultura S.A.

Reichel-Dolmatoff, G., 1985b. [1951] *Los Kogi* II. Bogotá: Procultura S.A.

Rival, L., 1993. 'The growth of family trees: Understanding Huaorani perceptions of the forest', *Man* 28 (4), 635–52.

Rival, L., 1998. 'Trees, from symbols of life and regeneration to political artefacts', in L. Rival (ed.), *The Social Life of Trees: Anthropological Perspectives on Tree Symbolism*. New York, NY: Routledge, pp. 1–36.

Sahlins, M., 2014. 'On the ontological scheme of beyond nature and culture', *HAU: Journal of Ethnographic Theory* 4(1), 281–90.

Simard, S., 2021. *Finding the Mother Tree: Discovering the Wisdom of the Forest*. New York, NY: Knopf.

Sturluson, S. and Anderson, B., 2015. *The Younger Edda Also called Snorre's Edda, or The Prose Edda*. Lavergne, TN: Create Space Independent Publishing Platform.

Viveiros de Castro, E., 1998. 'Cosmological deixis and Amerindian perspectivism', *The Journal of the Royal Anthropological Institute* 4(3), 469–88.

Viveiros de Castro, E., 2004. 'Perspectival anthropology and the method of controlled equivocation', *Tipití: Journal of the Society for the Anthropology of Lowland South America* 2(1), 3.

Notes

1. From January 2012 to November 2013.
2. The language of the Kogi, pertaining to the Chibchan linguistic family.
3. There is a Kogi creational story that illustrates the special relationship between trees and stones.
4. 'Paying' (*zɐbihi*) is a key Kogi practice or responsibility to reciprocate or 'feed' nature for what it gives and using its elements. Payments may be confessions, nourishing music and dance, body fluids, materials like seeds or cotton, certain mental exercises, coloured little stones, or 'imagining one gives food'. What, when, where and how the Kogi pay is guided by divination, and depends on the spiritual power and natural element that needs being paid.

5 COMPOSING WITH PLANTS
Discerning their Call
Julie Laplante and Kañaa

> calling, calling with, called by, calling as if the world mattered, calling out, going too far, going visiting.
>
> (Haraway 2015, 5)[1]

Introduction

Plants couple with wind. Inhabiting and regenerating the air with their scents, shapes, rhythms and expressions, evoking, provoking imaginations so powerfully it is hard to understand how they are so easily relegated to a silent insentient background.

The Aristotelian hierarchy of lives, which places plants on a lower scale on the basis that they do not have a nervous system, has endured through time. Following Taussig (1992, 2015, 9), we might argue that such a 'nervous system writing', or cognitive approach ignores plants' vital materialism (cf. Bennett 2010). Using such an approach, plant sentience has been relegated to the realm of fantasy, pushed to the background, despite the many characteristics and capacities of plants that suggest otherwise. Plants also are often deemed to be secretive (things to 'discover'). This is despite their vulnerability, as is exemplified by their outward expression as they synthesise molecules in the open air, which is precisely what scientists attempt to replicate in the laboratory or 'controlled environments'. Such attempts to control plants, however, are always surpassed by new indeterminacies, calling for other kinds of skills to tap into these potentials. If life solely proliferates through continuous adjustments, joining these unpredictable movements to know plants might be more beneficial then decomposing them. Even though Aristotelianism relegated plants to the bottom of the hierarchy of life, it still took into consideration their liminal position, describing them as a principle of universal animation and psyche (Coccia 2016, 21). Plants therefore are not intermediaries, agents of the cosmic threshold between the living and the non-living, spirit and matter – yet vegetal life remains a principle through which 'life belongs to all' (Coccia 2016, 22). I will thus argue that aromas, dreams, sonority and duration are where human, vegetal and elemental 'stirs' can occur (Laplante 2020).

From working with pajés (shamans), I have learned how certain plants are consulted for their wisdom and knowledge of the cosmos in Brazilian Amazonia (Laplante 2004). Likewise, with Xhosa *isangoma* (healers) in Cape Town, South Africa, I have learned that plants are a means to communicate with the ancestors, to purify dreams and are prized for their vitality, or ability to grow on their own, often in unexpected locations (Laplante 2015).

The plantation and the pharmacy follow complicit trajectories: whilst the former enslaves through monoculture and the latter extracts one or two molecules to be tested clinically. In both cases complexity is reduced for purposes of commodification and control. Deleuze and Guattari provide a way to understand this limitation connected to what they term arborescent thought, rather than rhizomic thought. They wonder 'whether plant life is not entirely rhizomatic' (1987, 6) and ask if thinking is linear or if the creeping rhizomic growth of grass better demonstrates how the roots and shoots of thoughts develop. Moreover, they suggest that the West has a special relation to the forest, and deforestation, carving out fields from the forest and populating them with seed plants. 'The East presents a different figure: a relation to the steppe or garden (or in some cases, the desert and the oasis)' (Deleuze and Guattari 1987, 18).[2]

Here, I do not want to contrast the West to the East, but instead will adhere to the rhizomic form, since it is both closer to plants' materialities as well as to the practices followed by the healers that I have encountered in fieldwork conducted on both sides of the Indian Ocean, in South Africa (Laplante 2015), in Java Indonesia (Laplante 2016), along the equator in Indigenous Amazonia (Laplante 2004) and more recently in Cameroon (Laplante and Kañaa 2020). More broadly, this approach is lateral and pertains to an opening towards other kinds of life rather than seeking to close in on them within a positivist paradigm, whether it be to identify, name, contain, measure or control.

In *Knowledge of Life*, Canguilhem retains Radl's interpretation that a scientist can either experience a filial sentiment – as belonging to nature – 'or he [sic] holds himself in front of nature as before a foreign, indefinable object' (2008, 63). The latter position has largely been favoured in science; here however, I want to explore the former as it enables one to understand both what healers do, as well as what scientists, in particular anthropologists, could do more.

Anchoring my thoughts mainly in fieldwork done with Kañaa, second author and healer in Cameroon Africa, we want to express ways in which one can be pulled towards specific plants, answering their call as well as calling upon them for guidance to attend to an ill-being. In this sideways sensorial approach, plants stretch, inhere, subsist and grow into the imagination and flesh, making vegetal materialities become human and vice versa, with possibilities 'to augment desire and effort to persevere in one's being' (Spinoza 2002, 122) for one and the other. Here we compose with plants by bringing them into presence in writing, doing what Taussig (2018, 245) calls 'wolfing moves', to displace anthropology's mode of writing by letting the plants (and animals) loose from the plantation, by re-enchanting.

> What if one became animal or plant through literature, which certainly does not mean literally?
> (Deleuze and Guattari 1987, 4)

Plants 'coincide with the forms they invent' (Coccia 2016, 25), releasing sound vibrations when opportune; 'musical form, right down to its ruptures and proliferations, is comparable to a weed, a rhizome' (Coccia 2016, 2). How might we compose this chapter along these lines?

The first section opens with what I am calling a vegetal line, one which escapes the botanical entity as the beginning point. The middle section, like the central panel of a triptych (within the broader triptych of the chapter), expresses ways of healing with plants by joining with a healer in Cameroon, which involves jumping from one vegetal line to another – sideways, as grass grows – and paying attention laterally. The final section offers evanescent echoes of other such sonorous yet silenced vegetal practices.

Elusive vegetal – beyond the botanical object

The main Linnean mode of classification of plants in science, which emerged in the eighteenth century, begins with the plant itself.

'*Potency* and *use* provide definitions that are worthless to a botanist' wrote Linneus in *Philosophia Botanica*, his treatise laying out 'the Science of Botany' (Hartigan 2017, 46). Following this dismissal of the verb to privilege the noun, or fixing differences in genus, loses sight of the necessary creative potentials that are needed for life to prosper; namely action and potential, which are not already there or are yet to come. For Coccia (2016, 146), plants and their structure are better understood through cosmology rather than through botany, since they are atmosphere. Beginning with plants as a discrete entity relies on analysis that has contributed to establish them as objects and is thus uninterested or unable to discern their call or their potency – what Chudakova (2017) calls pharmacopoeisis; the necessary lively continuous adjustments and improvisations needed to make a plant medicinal. The latter quality is thus not 'already there', nor 'waiting to be discovered', yet it is there as potential to do something medicinal that we do not necessarily yet know.

Taking the vegetal by the middle approach – or jumping from one vegetal line to another, rhizomically – offers a more interesting path since, as with all living, life 'has neither a beginning nor end, but always a middle (milieu) from which it grows and which it overspills' (Deleuze and Guattari 1987, 21). This approach enables one to challenge the idea of the plant as a discrete entity, body or agent with specific properties that we need to uncover and in so doing 'think of ourselves not as *beings* but as *becomings* – that is, not as discrete and pre-formed entities but as trajectories of movement and growth' (Ingold 2013b, 8) always in the process of becoming something else, human and non-human. It can also be understood as tendencies towards which we lean, or what Kohn calls 'effortless efficacy', 'how form emerges and propagates in the forest and in the lives of those who relate to it' (2013, 20). Kohn invites us to take seriously how forests think and, in some ways, how to think along with them in a sort of lively semiosis, yet his corresponding Peircean framework appears less useful here than a Spinozist one, which insists that we do not know what a body can do. Also taking this as a starting point, and adhering to this, Myers suggests anthropology is 'becoming transductor in a field of affects' (2019, 97) – affects here are understood as what is yet to emerge in an encounter or in-between. 'Affect is not the passage from one lived state to another, yet it is the non-human becoming of man [sic]' (Deleuze and Guattari 2005, 173)[3]. In other words, encounters with a vegetal, human, sonorous, animal, elemental, energetic body is not good or bad in itself, yet can create something new in-between, which can either augment or diminish, favour or inhibit our power to act (Spinoza 2002, 82). Plants offer a line of flight elsewhere, which I aim to draw by passing through some current blockages or stoppages; namely it undoes an idea of 'agency' that suggests it pertains to discrete entities, whether human or nonhuman (cf. Bennett 2010).

In anthropology there is a relative consensus that agency is 'power to act' as per Spinoza. Dissension however arises when it comes to qualifying or allocating it, often unevenly to humans or nonhumans, or by opposing it to the notion of structure upon which it seems to depend[4]. The notion of agency is generally undertheorised in anthropology, and it has mostly been allocated to humans. Even in the more-than-human turn that professes to level the agency playing field, just a little less agency is typically afforded to the non-human. The first way agency has been perceived is in a backward movement, or one that brings everything back to a pre-existing structure or a human agent. While Alfred Gell (1998) does not deal with plants directly, he suggests art (or things, the animal) can exercise social agency similar to the way social agency can be exercised in relation to things. For this we need to retrace the causal relation through cognitive operations enabling one to trace the object back to the human agent who would have given them their power to act. This position however disregards the possibility of plants having agency of their own, or one that passes through us as atmosphere, as we suggest here. Another classical work related to agency is by Judith Butler (2010), also primarily preoccupied by human agency and in a backward movement placing a subjectivation that precedes self-consciousness. From her studies on gender, Butler develops a notion of agency through a philosophy of action, which has been taken up in both anthropology and linguistics, and 'refers to the power to act of an individual submitted to a dominant power, which is not a will inherent in a supposedly autonomous subject but a capacity of an ontological and dialogical order' (Brunon 2015, 120). These approaches respectively maintain a certain porosity of the object and the subject, or a passing in-between, yet these are preceded correspondingly by a human agent or structure. There is no room for vegetal agency in our approach, as it transgresses both, undoing the agent by expressing structure in continuously novel ways.

A second approach allocates agency across the human and the nonhuman, yet still affirming a sort of distinct human agency. Drawing upon the work of Charles S. Peirce (1931–5), Eduardo Kohn for instance allocates iconic, indexical and symbolic semiosis to humans while the biotic nonhuman would be limited to the two first kinds of semiosis (2007, 6). The more interesting aspect with Kohn is the fluctuation of vital powers or agencies that can augment or diminish through time as noted with the Runa, also echoing Irving Hallowell (1960) in terms of Ojibwa ontology; these two authors stretch a notion of subject or of self to the nonhuman, including the vegetal, simply by reversing the subject-object dichotomy. This still does not seem to concur with the vegetal as it eludes the dichotomy altogether. Similarly, Bruno Latour's quasi-object or quasi-subject (borrowed from Michel Serres 1990), does not enable one to grasp the vegetal successfully. According to Latour '[t]here are at least two ways, one from semiotics and the other from ontology, to direct our attention to the common ground of agency before we let it bifurcate into what is animated and what is de-animated' (2014, 8). He critiques the notion that all the action is put in the antecedent, in the 'fact', the 'river' (or the 'agent') thus missing out on its eventfulness (2014, 14). Latour however retains the idea of the agent, suggesting that 'all the agents share the same shape-changing destiny', yet that the crucial political task is to 'distribute agency as far and in as *differentiated* a way as possible – until, that is, we have thoroughly lost any relation between those two concepts of object and subject that are of no interest any more except patrimonial' (2014, 17, emphasis in original). An agent however needs to do this distribution, a task that Latour gives implicitly to the human. In this quasi-animated movement of distribution of agencies, we can also include Jane Bennett (2010) and her 'agency of assemblages', which sees in the work of Deleuze a connection between human agency and some forms of nonhuman agencies yet affirms nevertheless a sort of distinct intentional human agency (Bowden 2015, 62). She evokes the active role of nonhuman materialities in public life as a not-quite-human capaciousness (vibrant matter), a thing-power, to which she nevertheless aims to give a voice (Bennett 2010, 3); in doing so, she makes herself at the same time spokesperson for this thing-power, as well as retains action in the hands/voice of the human. Vegetal power to act is thus kept minimal in these approaches.

There is another approach to the vegetal (indeed to reality). One that does not rely on the subject/object dichotomy, and in so doing, allows plants to be seen as expressions in the field of possibilities, part of 'a multitude of hummings [...] These vibrations of sight and sound, music and color, are turnings that to some people appear unpredictable, ephemeral, and may make you frighteningly vulnerable' (Taussig 2015, 33). Yet these 'turnings' endure and by following them, emulating, joining their speeds and slowness, intensities and thresholds as they pass through or become, laterally provides another valuable perspective on life living. This is where we can comprehend that a plant *is* its expression, aroma, texture aesthetic and taste, very much as how Ingold (2015, 82) suggests we understand that the glacier *is* its sonority, luminosity and palpability. He offers a notion of agency that comes from action (Ingold 2015, 124), from life lived with others attentionally rather than intentionally (Ingold 2015, 152). He maintains life in what Deleuze and Guattari (1980) call smooth space – in the matter flow of a world without objects, eclipses the possibility of agency as residing in objects, a gesture similar to Gregory Bateson's objection to the subject. Already in the 1970s Bateson concluded that the 'conscious intention' taken as a criteria of mental agency was a major error; namely

> that the idea of conscious intention [in the mind of social scientists] is artificial, an artifact or an epiphenomenon, a collateral product of a disastrous process in the history of western thought, and in fact, a pathogenic premise
>
> (Bateson 1991, 228)

especially as it is made to coincide on a concept of individual self. We thus need to dissolve the 'agent' to maintain the vital flow or power to act beyond the bounds of organic life.

For Massumi (2002), the real agency is essentially nonhuman, hence ontologically identified with something like the 'virtual' ground of all actual things and the events attributed to it. Simondon (1995) also signals to a sort of immanent or elusive energetic agency in becoming, moving from metastability to metastability. Strathern (1991) and Tsing (2014) also allude to relational entanglements which do their own distributing' beyond or across both an idea of agent, individuate or structure.

Hence while a plant is largely understood in a study of beings, it may be more accessible in a study of becomings – a dance of animacy, as suggested by Ingold (2013a, 100), which constitutes an effort to project ourselves in the world, one which is in the air, vegetal, human, mineral, animal … all mixing and stirring.

Discerning the call is to become more planty, to vegetalise our all-too-human sensorium as suggested by Myers (2017, 2019), and to learn to conspire with plants.

> In principle, *becoming-plant* would involve our extension and ideas entering into composition with *something else* in such a way that the particles emitted *from* the aggregate thus composed will *verb vegetally* as a function of the relation of movement and rest, or of molecular proximity, in which they can enter.
>
> (Houle 2011, 97)

This is a way that enables one to bring plants into academic compositions without becoming their spokesperson or turning them into objects. It seeks to notice attentionally *from* an increased awareness to their expressions in intimate copresence, very much as I recently learned with a healer in Cameroon, to whose practice we now turn, turning it into a composition which emerged from the encounter.

Kañaa's Healing Triptych – Stems from forests and leaves

> *On demande à la feuille qui porte la plus grande adversité.* (We ask the leaf that carries the greatest adversity).
>
> (Kañaa, personal communication, November 2020)

Kañaa is a Basaa healer of Bantu descent and founder of the *Association of Research in the Anthropology of traditional Medicine* (ARAM) of Etoa, in the periphery of Yaoundé, Cameroon.[5] The ancestral forest of Bassinglègè, from which he makes medicine, are some 100 km away from the headquarters in the direction of Douala where the Association has an Antenna (base) in Lamal-Pouguè. There, they have recently replanted fourteen hectares of deforested land with some of its indigenous medicinal plants. The access to the ancestral and replanted forests is secured through an arrangement with the Cameroonian Ministry of Forestry and Fauna and extends the possibilities of healing at the headquarters in Etoa, where people come through in a continuous flow. I was invited to give a conference there in August 2018, staying on the premises for over a month and we have since co-offered a fieldwork course twice (online due to the pandemic), in August 2020 and in June 2022.

The composition presented here emerged from a concerted effort to 'sound' (pay attention to rhythms, speeds and slowness in entangling with plants in healing) as a way to best express its efficaciousness. Before travelling to visit ARAM in the summer of 2018, and in preparation to work through sound, I followed a four-day Sonic Triptych Workshop; *A Sound Laboratory in Three Counterpoints* offered by Cubero, Herrera and Labelle (2018), where three facilitators demonstrated ensounding as a three part process, namely focused on the past, present and foreseeable future[6]. My fieldwork in Cameroon was further done with my 19 year old son and 17 year old daughter (again making us three), each of us in turn playing one or another rhythm in encounters, all of these triptychs making this composition in the form which follows. An accompanying soundtrack[7] expresses the three spatio-temporal figurations, hopefully creating sonorous sensations (Laplante 2020) or ways to discern plants at multiple levels as sense or event. It can be listened to before, after or along with the following image and text. The three spatio-temporal figurations here presented vertically in text are best seen horizontally as in the image below, with the possibility of either event becoming vibration, resonance or attendant interchangeably. Together the events, as told, point towards the ephemeral and elusive vegetal, which can nevertheless endure.

Composing with plants across these three spatio-temporal figurations evoke the headquarters in Etoa and their extensions in the replanted and ancestral forests respectively. We aim to express fluctuations in the intensity of human-plant relations in healing practices as matters of calling, being called and discernment. Upon exploring the different speeds and slowness, rhythms enacted in the practices of the three sites came to correspond to ways of creating sonorous sensations (Laplante 2020), in which plants, within other elements or life forms, took part. The triptych can further be understood through Deleuze's assessment of Francis Bacon's paintings, many of which are done in three horizontal figures giving three fundamental rhythms: active (simple sensation or vibration), passive (coupling of sensation or resonance) and attendant.

> With the triptych, finally, rhythm takes on an extraordinary amplitude in a *forced movement* that gives it an autonomy and produces in us the impression of Time: the limits of sensation are broken, exceeded in all directions; the Figures are lifted up, or thrown in the air, placed upon aerial riggings from which they suddenly fall. But at the same time, in this immobile fall, the strangest phenomenon of recomposition or redistribution is produced, for it is the rhythm itself that becomes sensation; it is rhythm that becomes Figure, according to its own separated directions, the active, the passive and the attendant.
>
> (Deleuze 2003, 60–1)

This is explained as a means of rendering Time sensible, which is a task common to the painter, the musician and sometimes the writer; and here I likewise attempt in my anthropological writing to suggest the healer is, in this case, also accomplishing this task to heal. More specifically, the healing done in the ARAM headquarters occurs through the intense vibrations of people and things passing through, which are understood to tap into different kinds of life by loosening them up in passing (such as by surprise or through playful performances). This animated or active site seems to provide a simple sensation or vibration on its own, which is strengthened by evoking the forest both in cosmology and materiality.

The first forest is fragile and only recently replanted, and is somewhat passive as all await its growth; yet it resonates with the headquarters or first site, and offers a counterpoint as it provides a place to call upon plants and attune to their calling in search of remedies. Finally, the ancestral forest takes all the former practices to another level of cosmological force, transcending space and time and is able to provide joy, as well as sadness if not dealt with carefully. It enlivens all of the other practices, giving them depth and wisdom in both Time and Space. 'It is a task beyond all measure or cadence' (Deleuze 2003, 54).

PLANTS MATTER 119

Fig. 5.1 TIME / RESONANCE / VIBRATION

It is a nonnarrative, nonlogical and most of all the triptych avoids thinking in dichotomies or in causal logic; it is nonlinear, favouring circularity, as well as being horizontal, with every spatio-temporal figuration potentially playing any of the three rhythms.

Vibration-intensity-active rhythm

ARAM's headquarters is where Kañaa lives and hosts patients on a daily basis. The intensity of the constant flow of people arriving with fresh plants, water, bread, palm wine as gifts is remarkable. Similarly, he is remarkable, as he walks barefoot – seemingly improvising – across the vegetal, animal, human and elemental to heal in surprising ways, loosening tensions, unblocking passages. Upon arrival and staying on the premises for more than a month in the summer of 2018, it felt as if I were entering what Bateson (1972) calls a plateau, or a region of intensity vibrating on itself, with all kept alive, yet with nothing reified. Plants grow semi-wild in different locations of the land. The plants collected in the forests near the Antenna are dried, cut, prepared and classified in a small house on the parcel. They might be cooked into different meals, made ready for healing purposes, or they are prepared into creams for varying massage treatments. While these activities with plants are mundane and part of everyday practices, they also engage, stretch and extend their potentials through both time and space, as numerous treatments take place at a distance, long after the patient has left. The beer bottle in the left image above, for instance, is still working on a client who has problems with alcoholism, and we need to beware of displacing it. The white rooster also takes part in numerous treatments, sometimes absorbing someone else's injury, today, for instance, missing a leg in replacement for the healer's injury to his foot.

The facility, as shown in the image, is divided between a divining room where some of the massages occur, and a kitchen on a ground floor. Other spontaneous treatments can occur outside, in the living room of the main house or in the kitchen as part of a broader performance that might aim to loosen up a problem. Most treatments however happen around the fire at the entrance of the headquarters. Some involve all who are present, sometimes as meals, or discussions and also as playful performances where, for example, the healer may wake up in the morning in the role of a silent mime and remain as such throughout the healing sessions of that day right up until nightfall. When in character he enacts caricatures and employs techniques that both amuse and shock. This might be by imitating people's gestures to perfection, by emptying one client's purse, or by insisting that another remained seated with some fruits and vegetables on his head and knees until he found courage. These performances stimulate healing by undoing rigidities often attributed to adhering to institutionalised religions, which are deemed to clutter issues within its mesh, and obstruct flows. Plants take part in all of these performances.

Resonance-calling-corresponding passive rhythm

Kañaa is always alert and aware of the plants calling as he searches for the remedy to a patient's illbeing; plants make themselves visible as we meander in a young recently replanted forest near the Lamal-Pouguè Antenna. Walking leisurely in a straight line for a while identifying, naming and tasting a variety of different fruit, roots, seeds, leaves and barks along the way, we pass through a cluster of spectacular young cocoa trees enchanting with their purplish-orange-pink fruit at all heights of the trunk. The rhythm of the walk drastically shifts into a different speed and attention. Very abruptly, Kañaa changes mode, starts moving in odd and unexpected directions, attentive as if hunting an animal. Suddenly he plunges into the forest, stretching his arm out asking for his machete with a sense of urgency. He kneels at the foot of the young tree, pauses resting his forehead on one side of its trunk and taps it with his hand on the other side. Slowly he then places the blade of the machete horizontally along the trunk, slides it down delicately and collects a few thin slivers of the bark which we place in the large *Hicoma* leaf, which serves as an envelope, and he begins to meander in another direction, attracted towards a second tree; he explains:

> One plant is calling upon another plant, and the order is essential.

Another type of plant materiality (or vitality?) is necessary to homogenise the bark and any other leaves we collect. It is either my daughter, son or I who needs to collect this plant as it needs to correspond to the combination of our particular bodily assemblage or constitution, as we all come from the same place and household in Canada. Our correspondence with the vegetal is required since the remedy is to help straighten my son's back; part of a treatment that has already begun through massage at the headquarters in Yaoundé. My daughter nudges me to collect the leaves so I move towards the indicated plant, mesmerised by a red-pink-purple liana that was twirling around the young tree. I slowly pick the nine leaves I was asked to collect, which we then place carefully in the same large leaf of the *Hicoma* tree where the other plants are also kept. None of these plants are given names in the moment, yet when I asked the name of the latter plant a few years later, I am given nine names: *Lum; Djadian; Ibogi dodogi; Lom évong; Teng: Totom; Ndodong; Bawai (ben); et Titimut*. While they are all from the *Djèe* (cocoa) tree, each leaf has a different name with relation to their order of attraction, which corresponds to a different action they will provide in the 'mélange'. Along the way, later in the treatment, the peel of a green pineapple will also be added to the mixture. However, a final plant, situated in the ancestral forest of Bassinglègè will be the last ingredient. This will not act to loosen the back so as to make it straight, but rather to solidify it now that it has been straightened and ensure that circulation is flowing again. After acquiring the necessary ingredients for this initial mixture, we, following the same mode, searched for three further combinations of plants for three other cases that required treatment before returning to the Antenna and headquarters.

Time-joy-attendant rhythm

Agitated, nervous, slightly larger than the others as it indulges itself with too much oxygen; it easily attracts attention amidst its multiple peers. It is telling (announcing, yelling out?) that it is ready to do something else. On the verge of taking a line of flight, it had already taken on rhythms of its own for a while, and it is now on the point of becoming something else, ready to leave the young tree from which it sprung. Before falling to join other lively motions of decomposition or composting. It offers its insights or wisdom and can be intercepted by a healer to make us immediately invisible so as to tap into foreseeable futures.

Once such an agitated leaf from a nearby young tree was collected and presented to me, Kañaa explained:

> 'You have been invisible, we have worked on this earlier' he says while looking at the two guides to receive their acquiescence. He continues: 'Now that you are made visible again, we are able to see if what you are doing here will succeed, in terms of local support with ARAM as well as support from colleagues in Canada, which in turn will decide if we continue walking deeper into the forest or if we need to turn back'.

The leaf is then placed on my forehead, and we watch how it falls.[8] It falls on its downside, indicating obstacles ahead and the impossibility to continue our walk. Another leaf calls attention and we repeat the same scenario until the seventh leaf finally falls on its upside on my foot. Where the leaf falls is also indicative, in this case confirming our work will be fruitful and we can continue walking into the forest. What is made visible by the leaf penetrates past the immediate and provides the potential to take the past and future into consideration. This can also be done with coins, which offer other materialities through which to see further in space-time in immediacy.

While the leaves of young trees offer insights, ancestral trees of the Forest of Bassinglègè offer vitality, one with which one needs to get involved with carefully as it is highly ambivalent with potentials to both harm as well as enlighten. In August 2020, during the first online session of a fieldwork course with ARAM (turned online due to the pandemic situation), Kañaa appeared onscreen at the foot of an ancestral *Djab* tree in the forest of Bassinglègè with a patient and some other members of the association. The treatment offered to a young Cameroonian consists of letting the powers of the tree flow through him, explaining the possibility that the tree disavows him or does not heal him unless he activates himself on his side to re-establish good relations with his immediate family which were in turmoil. This affective ambivalence of the power of the tree caught the attention of numerous students from the University of Ottawa attending the course, especially a Cameroonian student who seemed to best grasp the intensity of such an engagement with a tree, which endures in space-time. She was most worried of the implications of becoming tied to a tree, seemingly aware of its real potencies, as well as the possibilities for the healer to accomplish this passage of potentialities from the tree to the human and which Kañaa explains in the following manner,

> In the case of an energetic transfer between a plant and a human, the specificity resides in the neutrality and the position of the person who effectuates this work. The practician, who plays an essential role in this transfer, is an element who speaks with terrestrial and human nature.

This 'neutral' point is also described as becoming nimble, light and agile in movement, open to vitalities which can pass through him yet can also be passed on to others. Kañaa performs this suppleness at all times, always barefoot, whether on earth or on polished building floors; he avoids any authoritarian position, always available and alert to potentials that enable healing by unblocking or adjusting the passage of vitalities across the living. He generally avoids touching people, but, if he does, it is a spontaneous act, designed to unblock a knot or tension point that he senses. He releases the knot to encourage the 'passage of vitalities' by the touch of his hand at a key place on the body at a moment when the person is not expecting it. Should he touch someone inadvertently, he does a gesture above their heads to release the excess power, which can otherwise give a bad headache. During massages, he sometimes signals his assistant to move away as he completes his rhythmic movement which accelerates and is accompanied by profound expirations and strident screams that signal the end of the passage of powers.

The kind of body that is attended to is thus this body, open to other kinds of bodies, one which is non-anthropocentric and composed of multiple bodies as per Spinoza's notion; they are *bodies* that distinguish themselves in terms of their potential movements and rest, speeds and slowness (modal distinctions between affected parts of diverse manners) and not in terms of their substance which is indivisible (Spinoza 2002, 75). This is what Deleuze and Guattari (1980) call, following Antonin Artaud (1934), the Body without Organs (BwO), or the body which is not defined by the organism and which we need 'to keep just enough for it to reform itself at each dawn' (Deleuze and Guattari 1980, 199). The BwO is thus an assemblage which does and undoes itself continuously in connection with other assemblages; it is a supple, porous body – open to infinite bodies, able to enter in correspondence with other bodies. Understanding a plant as an assemblage enables one to grasp its versatility as it constitutes one of the loosest assemblages, able to let go of most of its parts while at the same time staying alive, even prospering in this manner, namely by coupling with wind, water, sun, sound, and also with the animal and the human. Becoming supple thus enables more ample potential correspondences with other assemblages.

These phenomena can be understood in the sense of deleuzoguattarien becomings[9] and here more specifically in the sense of becoming-plant (Houle 2011, Laplante 2016, 2017, 2020, Marder 2020), while the human forms a block (of becoming) with the vegetal in ways through which something new emerges in-between. According to Houle (2011, 112):

> Plant-becoming also radically re-imagines Life as that which can be accomplished not within a successfully-managed organic encasement of what a thing is (its being, its teloi, its progeny) but, as that which can happen by virtue of *a certain unfaithful power of connectivity*.

Attention thus turns towards what is on the point of becoming, or towards that which life tends.

> Becoming, after all, is a journey without a fixed destination and, even though something is supposed to come out of vegetal becoming (fruit and seeds are the obvious references), this 'something' lacks ultimacy, because it, in its turn (the turn of *vertere*), initiates a new growth and decay. Circular closure coincides with absolute openness.
> (Marder 2020, original English version)

At the very end of this walk which occurs two weeks after the one in the replanted forest, Kañaa notices a tall straight tree and collects some of its bark. The inside of the bark, a reddish-brown colour like the earth, is soft and moist yet so fine that it appears as a dry powder once we roll it in the palms of our hands. Its finesse and suppleness offer the possibility to fix the back that is now straight, insuring the passage of vital flows. It is the final plant materiality that is added to the mixture. The augmentation of the effort and desire to persevere in existence is thus attended to by seeking to augment these vital circulations, which are referred to as the joy of Bassinglègè.

Evanescent echoes

A decade earlier I had spent seven years following a plant, opting for a low profile one, or at least one which was not yet sensational, only to realise how it had its own museum in Paris and houses of Artemisia were being built all over West Africa. It was however not the same Artemisia that I had followed; not the one that people drink and that was made illegal in Europe for its euphoric affects in the form of absinthe; not the one turned into a protocol biopharmaceutical against malaria for its extracted molecule of artemisinine 'discovered' in 1972 by Tu Youyou, a Chinese pharmaceutical chemist and malariologist who received the Nobel Laureate in physiology or medicine in 2015. It was another one; a plant growing wild through the sub-Saharan region set under the bedsheets to purify dreams before undergoing a healing session or initiation with Xhosa *isangoma* (healer of bantu tradition). Perhaps it had nothing to do with the family Artemisia or its subspecies; perhaps I was following something else that emerged in-between plants and people.

I learned of the plant as it was entering the process of turning into an object, species, commodity to test in a preclinical trial against mycobacterium tuberculosis (Laplante 2015). This trial from The International Center for Phytotherapy Studies (TICIPS pronounced 'tea-sips') was leading another trial with Sutherlandia frutescens against HIV/Aids and I was rather attracted by the more subtle preclinical trial of *Artemisia afra*, as well as its interest in a more subtle and long-term disease of the tropics, namely tuberculosis. What interested me most was how this clinical process was also proclaimed as a way to recognise traditional Indigenous knowledge. Although it began with the hint of a broadly used plant of the traditional lore, the preclinical trial had little to do with the Indigenous practices once it brought the plant into the laboratory. There, it had much more to do with the invention of a new entity that could be enclosed and exploited, even enslaved. This is how *isangoma*, as well as bossiedoktors (Rastafarian bushdoctors), found it to be when they saw it in ten cultivated rows to maximise its homogeneity in a big batch to be tested for the trial; in this state they had no interest for it, saying it had lost its life or efficacy equated with its vitality, and it is the latter which they were after (Laplante 2015, 2012, 2009a, 2009b). While this was known by the molecular biologists leading the preclinical trial, this was disregarded in their own quest to capture a molecule in time to catch a ride on the global health market which closes in on life to let it proliferate afterwards, instead of opening up to it directly.

These vegetal entanglements proliferating life through life linked to the inherent power of plants, visible, audible or felt through vibrations are also alluded to by Ju/'hoan healers in Namibia, who let themselves be attracted by them:

> Medicinal plants induce a response in humans through their inherent potency, their own knowledge/knowledgeability, and their smell (tsà'á) or essence moving through the air, the wind and in breath.
>
> (Gibson 2018, 6)

As Low (2007, S75) notes for the Khoisan of Southern Africa,

> wind or smells and pheromones draw and repulse organisms to their mutual ends. When the wind of an animal or plant enters a Khoisan body there is a unity between the two phenomena. Wind smells lock participants into a web of relationships. The essence of one organism connects with another.

Current bossiedoktors in South Africa seek their knowledge from the latter, yet this is also common knowledge that is not limited to healers when a plant manifests its presence while seeking a remedy; for instance, a woman from Matzikama in the Western Cape, South Africa, explains,'That bulb was telling/signalling to me, "See, use me!" [hy't my gesein, Sien, gebruik my!' (Gibson 2018, 7); a Rastafari herbalist explains that the 'spirit [gees] of the plant "pulls you back" [*trek jou terug*]' (see Nathen 2016). It's a question of 'being attentive' (Nathen 2018) to the particular expression of a singular plant as it attracts attention, sometimes to lead towards another plant or directly indicating its power to act. The question of vitalities or flows of energy (Cohen 2015 in Green 2020, 97) is immanent to the vegetal, yet also to the human.

> If the orientation to health includes a wider array of toxins and taxonomies that contribute to the experience of having energy or vitality (*krag* or the different, although not entirely dissimilar concept of 'qi'),[10] – then biochemical research need not necessarily begin with the particular pathway of seeking compounds related to pathogen elimination.
> (Green et al. 2015, 9)

A beneficial treatment then emerges with the ability to find the right assemblage or the fortuitous combination in terms of flows and vitalities (having less to do with species' identification or extraction of molecules), and this is done through constant minute adjustments to ensure the proper correspondences; a process which consists in calling upon plants as well as listening to their call.

In this current pandemic derailment, many are looking towards the vegetal. The vegetal emerges mostly as isolated molecules synthesised, unrecognisable, as part of a global biomedical tradition, already having lost sight of the plant from which they came. Plants however don't appear as lawful allies, as ways of enlivening bodies and regenerating the air we breathe, perhaps also reviving our cosmology. This calls to deal with different kinds of bodies. While the USA ran out of toilet paper and turned to buying guns to care for bodies when the pandemic hit, Indonesia ran out of Jamu, which had also increased in consumption during avian flu in 2003 (Dixit 2020). The Javanese ayurvedic tradition of Jamu specialises in ensuring bodily fluids flow to the right intensities, speeds and slowness, enlivening through vegetal flows of hundreds of spices, herbs, barks, all prepared in slow back and forth rhythms to obtain the right consistency and density seen to fit the times. All the women preparing Jamu do so in a slow motion of entering into the plant and leaning back, extracting their juices to make drinks that will enliven, unblock or offer a chance encounter (Laplante 2015a, 2016). Might this be what Nietzsche meant by 'One must still have *chaos* in oneself to be able to give birth to a *dancing star*'?[11] Amid the chaos of being, Nietzsche believed that plants offer us inspiration for living.[12] In Java, as in Cameroon, they literally offer this vitality.

I have argued for an approach to the vegetal in suspense. One that requires an agility to discern plants' calling expressed in the very air we breathe, mix and compose with. I've suggested that to do so we need to move through and past what can be seen as stoppages and blockages, whether they be ideas of agents, agency or discrete bodies, species or botanical entities to identify, extract or control. The healing triptych aims to express how this occurs by jumping linearly on vegetal lines. The final echoes note that this is done elsewhere and might be something that can be done more in science, especially in anthropology, both in practice and in writing.

Epilogue: Composing with plants; discerning their call

Wandering in the ancestral forest of Bassinglègè with a Bantu healer in Cameroon, he lets plants attract his attention to discern how to heal someone who had earlier sought his guidance. Thinking with plants, other living materialities, human and not, is his vocation or his calling. A calling, as in calling with, called by, calling as if the world mattered, can be further understood as an inner impulse leading us on a particular path, in this case one inspired by what the world has to tell or how it manifests its presence. This filial sentiment of belonging and surrender can be contrasted with most of today's ways of doing science in which the researcher holds himself in front of the world or of 'nature' as before a foreign, indefinable object. I explore the former position as it enables us to understand what healers do, as well as what scientists could do more, especially anthropologists. This approach moves past an idea of 'agency', which corresponds with what Ingold refers to as 'knowing from the inside' (2013a, 1), an expression also used in von Goethe's *The Metamorphosis of Plants* (1790) where he proposed a form of delicate empiricism. I further borrow from Spinoza, Bateson and Deleuze's notion of bodies open to one another through affective flows passing from one into the other across a plane of immanence. Thinking of ourselves as an assemblage (agencement) of multiple bodies in continuous becoming, which can correspond with other assemblages in more or less attuned and beneficial ways, helps make sense of ways plants pass through us as well as how they transform us.

References

Arendt, H., 1982. *Lectures on Kant's Political Philosophy*. Brighton: Harvester.

Artaud, A., 1934. *Héliogabale ou l'anarchiste couronné*. Paris: Denoël et Steele.

Bateson, G., 1972. *Steps to an Ecology of Mind*. New York, NY: Balantine Books.

Bateson, G., 1991. *Sacred Unity: Further Steps to an Ecology of Mind*. New York, NY: HarperCollins.

Bennett, J., 2010. *Vibrant Matter: A Political Ecology of Things*. Durham, NC and London: Duke University Press.

Butler, J., 2010. 'Performative agency', *Journal of Cultural Economy* 3(2), 147–61.

Brunon, H., 2015. 'L'agentivité des plantes', *Vacarme* 73, 118–23.

Bowden, S., 2015. 'Human and nonhuman agency', in J. Roffe and H. Stark (eds), *Deleuze and the Non/Human*. Hampshire: Palgrave Macmillan, pp. 60–80.

Canguilhem, G., 2008. *Knowledge of Life*. New York, NY: Fordham University Press.

Chudakova, T., 2017. 'Plant matters: Buddhist medicine and economies of attention in postsocialist Siberia', *American Ethnologist* 44(2), 341–54.

Coccia, E., 2016. *La vie des plantes. Une métaphysique du mélange*. Paris: Éditions Payot and Rivages.

Cohen, J., 2015. 'Kruiedocters, Plants and Molecules: Relations of Power, Wind, and Mater in Namaqualand' (Unpublished PhD thesis, University of Cape Town).

Cubero, C. A., Herrera, P. D. and Labelle, B. 2018. 'Sonic triptych: A sound laboratory in three counterpoints'. CASCA-CUBA annual meeting, Santiago de Cuba, 16–20 May. Cuba: CASCA and SfAA.

Deleuze, G., 2003 [1981]. *Francis Bacon: The Logic of Sensation*. New York, NY: Continuum.

Deleuze, G. and Guattari, F., 1980. *Mille plateaux. Capitalisme et schizophrénie 2*. Paris: Minuit.

Deleuze, G. and Guattari, F., 1987. *A Thousand Plateaus: Capitalism and Schizophrenia*, trans. B. Massumi. Minneapolis, MN: University of Minnesota Press.

Deleuze, G. and Guattari, F., [1991] 2005. *Qu'est-ce que la philosophie ?* Paris: Minuit.

Dixit, P., 2020. 'Guns in America, Jamu in Indonesia. Panic buying during Covid-19, its privilege and what it forgets', *The Guardian*. [online] Available at: *https://www.theguardian.pe.ca/opinion/ regional-perspectives/prajwala-dixit-guns-in-america-jamu-in-indonesi a-panic-buying-during-covid-19-its-privilege-and-what-it-forgets-427383/*. Accessed Sepember 2021.

Farquhar, J. B., 2012. 'Knowledge in translation: Global science, local things' in L. Green and S. Levine (eds), *Medicine and the Politics of Knowledge*. Cape Town: HSRC Press, pp. 153–70.

Gell, A., 1998., *Art and Agency. An Anthropological Theory*. Oxford: Clarendon Press.

Gibson, D., 2018. 'Rethinking medicinal plants and plant medicines', *Anthropology Southern Africa* 41(1), 1–14, doi: 10.1080/23323256.2017.1415154.

Green, L., 2020. *Rock | Water | Life: Ecology and Humanities for a Decolonial South Africa*. Durham, NC and London: Duke University Press.

Green, L., Gammon, D. W., Hoffman, M. T., Cohen, J., Hilgart, A., Morrell, R. G. et al., 2015. 'Plants, people and health: Three disciplines at work in Namaqualand', *South African Journal of Science* 111(9–10), 1–12, doi: 10.17159/sajs.2015/20140276.

Hallowell, A. I., 1960. 'Ojibwa ontology, behavior and world view', in S. Diamond (ed.), *Culture in History: Essays in Honor of Paul Radin*. New York, NY: Columbia University Press, pp. 19–49.

Haraway, D., 2015. 'A curious practice', *Angelaki: Journal of the Theoretical Humanities* 20(2), 5–14.

Hartigan Jr, J., 2017. *Care of the Species: Races of Corn and the Science of Plant Biodiversity*. Minneapolis, MN: University of Minnesota Press.

Haudricourt, A., 1962. 'Domestication des animaux, culture des plantes et traitement d'autrui', *L'Homme* 2(1), 40–50.

Haudricourt, A., 1964. 'Nature et culture dans la civilisation de l'igname: l'origine des clones et des clans', *L'Homme* 4(1), 93–104.

Houle, K. L. F., 2011. 'Animal, vegetable, mineral: Ethics as extension or becoming? The case of becoming-plant', *Journal for Critical Animal Studies* 9(1/2), 89–116.

Ingold, T., 2013a. *Making. Anthropology, Archeology, Art and Architecture*. London and New York, NY: Routledge.

Ingold, T., 2013b. 'Prospect', in T. Ingold and G. Palsson (eds), *Biosocial Becomings: Integrating Social and Biological Anthropology*. New York, Cambridge University Press. pp. 1–21.

Ingold, T., 2015. *The Life of Lines*. London and New York, NY: Routledge.

Kañaa, R. A., 2018. *Médecine traditionnelle et savoirs thérapeutiques endogènes*. Paris: L'Harmattan.

Kohn, E., 2007. 'How dogs dream. Amazonian natures and the politics of transspecies engagement', *American Ethnologist* 34(1), 1–22.

Kohn, E., 2013. *How Forests Think. Toward an Anthropology Beyond the Human*. Berkeley, CA, Los Angeles, CA and London: University of California Press.

Laplante, J., 2004. *Pouvoir guérir. Médecines humanitaires et autochtones*. Québec: PUL.

Laplante, J., 2009a. 'South African Roots towards Global Knowledge: Music or Molecules?', *Anthropology Southern Africa* 32(1–2), 8–17.

Laplante, J., 2009b. 'Plantes médicinales, savoirs et société: vue des rastafari sud africains', *Drogues, santé et sociétés* 8(1), 93–121.

Laplante, J., 2012. '"Art de dire" Rastafari: créativité musicale et *dagga* dans les *townships* sud-africains', *Drogues, santé et sociétés* 11(1), 90–106.

Laplante, J., 2015. *Healing Roots. Anthropology in Life and Medicine* (Vol. 15). Oxford and New York, NY: Berghahn Books.

Laplante, J., 2015a. *Jamu Stories*. Anthropological Film 104 minutes, mis en ligne le 5 mai 2013. Available at: https://www.youtube.com/watch?v=CMRZRw1z2Fw. Accessed, 12 November 2020.

Laplante, J., 2016. 'Becoming-plant: Jamu in Java, Indonesia', in L. Olson and J. R. Stepp (eds), *Plants and Health: New Perspectives on the Health-Environment-Plant Nexus*. Cham: Springer International Publishing, pp. 17–65.

Laplante, J., 2017. 'Devenir-plante: enlacements vivants en Océan Indien et en Amazonie', *Drogues, santé et société* 16(2), 36–54.

Laplante, J., 2020. 'Sonorous sensations: Plant, people and elemental stirs in healing', in J. Laplante, A. Gandsman and W. Scobie (eds), *Search After Method. Sensing, Moving, and Imagining in Anthropological Fieldwork*. Oxford and New York, NY: Berghahn Books, 21–48.

Laplante, J., 2021. 'Agentivité', in *Anthropen.org*, Paris: Éditions des archives contemporaines, doi:10.47854/NJFW6857.

Laplante, J. and Kañaa, 2020. 'Appel des plantes. Joie de la forêt de Bassinglègè, Cameroun', *Anthropologie et sociétés* 44(3), 171–94.

Latour, B., 2014. 'Agency at the time of the Anthropocene', *New Literary History* 45, 1–18.

Low, C. 2007. 'Khoisan wind: hunting and healing', *The Journal of the Royal Anthropological Institute*, vol. 13, pp. S71–S90.

Marder, M. 2020. 'Vertimus. Dix thèses sur le devenir-plante', *Anthropologie et sociétés* 44(3), 195–206

Massumi, B., 2002. *Parables for the Virtual: Movement, Affect, Sensation*. Durham, NC: Duke University Press.

Myers, N., 2017. 'From the anthropocene to the plantropocene. Designing gardens for plant/people involution', *History and Anthropology* 28(3), 297–301.

Myers, N., 2019. 'Anthropologist as transducer in a field of affects', in N. Loveless (ed.), *Knots & Knowing. Methodologies and Ecologies in Research-Creation*. Edmonton: University of Alberta Press, pp. 97–126.

Nathen, T., 2016. 'Exploring "Assemblages:" A Multispecies Ethnography of the Relationship between Plants and People in the Gardens and Mountains of Klawer in the Matzikama Municipal Region, South Africa' (unpublished MA dissertation, University of the Western Cape).

Nathen, T., 2018. '"Being attentive": Exploring other-than-human agency in medicinal plants through everyday Rastafari plant practices', *Anthropology Southern Africa* 41(2), 115–26, doi: 10.1080/23323256.2018.1468720.

Peirce, C. S., 1931–5. *Collected Papers of Charles Sanders Peirce*. Cambridge, MA: Harvard University Press.

Serres, M., 1990. *Le Contrat naturel*. Paris: François Bourin.

Simondon, G., 1995. *L'individu et sa genèse physico-biologique*. Grenoble: Éditions Jérome Million.

Spinoza, B. de, 2002 [1849]. *L'éthique*, trad. Française de Saisset, version 1.0 PDF, by David Bosman, published June 4 2002. [online] Available at: *http://palimpsestes.fr/textes_philo/spinoza/ethique.pdf*. Accessed November 12 2020.

Strathern, M., 1991. *Partial Connections (ASAO Special Publication 3)*. Savage, MD: Rowman and Littlefield.

Taussig, M., 2018. *Palma Africana*. Chicago, IL and London: University of Chicago Press.

Taussig, M., 2015. *The Corn Wolf*. Chicago, IL and London: University of Chicago Press.

Taussig, M., 1992. *Nervous System*. New York, NY and London: Routledge.

Tsing, A. L., 2014. 'Strathern beyond the human: Testimony of a spore', *Theory, Culture and Society* 31(2/3), 221–41.

Notes

1. Quote from Haraway (2015, 5) referring to Vincianne Despret's work as well as alluding to Hannah Arendt's suggestion to train 'one's imagination to go visiting' (1982, 43).
2. André Haudricourt suggests this as the opposition between the moralities or philosophies of transcendence dear to the West and the immanent ones of the East (1962, 1964).
3. Translations from French are my own.
4. See the Anthropen dictionary entry 'Agentivité/agency' (Laplante 2021).
5. Aram was founded in 2011 and information can be found on its website *https://arametra.org/*. See Kañaa 2018, also Laplante and Kañaa (2020) for a more in-depth account of his practices.
6. CASCA/Cuba conference, Santiago, Cuba May 2018. In this experiment, I partook to ensounding an abandoned coffee plantation, giving a sense of the past (in the present).
7. *Rainforest Abstractions* field recordings by Oscar Robertson. Composed and edited by Marlie Robertson, *https://www.youtube.com/watch?v=Ef5CguBqqhY*.
8. To visualize this scene, see the *Anthropologie et sociétés* (co-ed. Laplante and Brunois-Pasina 2020) special issue video: *Devenir-plante. Enlacement et attachements*, at minute 5:50, *https://www.youtube.com/watch?v=mfE1yZC-nx0*.
9. Deleuze and Guattari (1980, 17) show how the becoming-bee of the orchid and the becoming-orchid of the bee adds plus-value for one and the other in-between.
10. See Cohen (2015) on concept of krag and Farquhar (2012) on the concept of 'qi'.
11. Friedrich Nietzsche. *Thus Spoke Zarathustra*, Prologue, 5[2].
12. *https://aeon.co/videos/amid-the-chaos-of-being-nietzsche-believed-that-plants-offer-us-inspiration-for-living*.

6 THE MATTER OF KNOWING PLANT MEDICINE AS ECOLOGY
From Vegetal Philosophy and Plant Science to Tea Tasting in the Anthropocene

Guy Waddell

For Minnie and Bea

Introduction

Although having antecedents as far back as 1938 in Vernadsky's 'noosphere' (Vernadsky 1998; Weart 2008), the term 'Anthropocene' came to popular attention in 2000 when coined by the atmospheric chemist, Paul Crutzen (Schwagerl 2014). The designation of Anthropocene asserts that human impact on the planet's functioning now warrants a distinct geological epoch, in which anthropogenic climate change and effect on ecologies sit centre stage. The biosphere, atmosphere, oceans and cryosphere are all now dependent on what humans do next. Over recent years, and in parallel to the irresistible pull of ideas orbiting and colliding with the Anthropocene, including the Plasticene, Chthulucene, Homogenocene, Capitalocene and Trumpocene (Mentz 2019), significant work has been undertaken in a number of disciplines to reconceptualise plants in ways that move humans from the epicentre of everything and allow for the agency of sessile critters. As well as a search for understanding, this plant-turn may be seen as an urge to ask for help from those who photosynthesise. Importantly, as will be argued, it is difficult, if not impossible, to think of plants without thinking of their ecologies. As such, plants offer a way in, even an invitation, to reworld the matter of the Anthropocene. This chapter takes embryology as a point of departure from which to consider how developments in vegetal philosophy, plant science and Attala's (2017) Edibility Approach may be worked with to explore how knowing medicinal plants differently, particularly through being medicinal via their phytochemical relationships with wider nature from which they are inseparable as well as through tasting and drinking them in, can result in human-plant ecologies with which we can live.

Matter matters

Embryology seems like a good place to start. As such, it reveals similarities between plants and human animals in very early life, during which the matter of plants and humans can meet in productive ways, and even make kin in troubling times (Haraway 2016).

Embryology describes the movement of matter in the process of formation. Matter matters in embryology and, along with anatomy, is arguably the most material of sciences. Both complex plants and animal bodies develop from spheres, which, as Houle (2019) points out, are maximally symmetrical objects that face all directions at once. While developing from either pollinated seeds or fertilised eggs, this spherical nature soon changes in the rush to become an object-body with 'directions'. Zernicka-Goetz (2020), in a compelling story of her life's work in human embryology, reveals the journey that led to the recognition that cellular symmetry-breaking is necessary for cells within the early human embryo to become different from neighbouring cells. It is this breaking of symmetry that allows cells to become something else. Her focus was to find the source of the symmetry breaking. Eventually this source was shown to be the anatomy of the cells themselves – where the location of the fatty nodular porous filamentous nucleus, the almost impossibly infolded mitochondria, the rough and the smooth endoplasmic reticulum or other organelles of varying shapes, textures and properties were shown to influence the fate of daughter cells. The distribution of matter in cells, matters. However, and importantly, this is also true for plants: plant cells proliferate by symmetric division while asymmetric division results in new cell types that arrive as the final form of the plant (Rasmussen et al. 2011; Pillitteri et al. 2016). Hence, in both human and plant embryology, the cells are not simply holding their contents, but function in very particular places to have very particular effects. This suggests that intracellular substances arrange themselves in subtle ways that determine the processes of differentiation, even if 'difference' is limited by the properties of these materials.

Split critters sensing open beings

Early life reveals many differences between plants and human animals. For example, plants erupt in growth, their cells differentiate quickly, and they grow from their periphery. On the other hand, animals, including human animals, are, at first, both slower to grow and to manifest cellular differentiation and grow to a limited but somewhat variable size and remain there. Houle (2019) points out that while both plant and animal embryos have vertical and radial axes of development, the integration of these axes in plants does not lead the plant to the same inwardness and outwardness that humans have. Rather than looking at axes of development, Pouteau (2019) looks at axes of symmetry to suggest another dissonance between plants and humans. Most animals, including humans, have three axes of symmetry: vertical, horizontal and sagittal which, respectively, have oppositional pairings of top/bottom, left/right, and front/back. This makes it difficult for human animals to understand what it is like for a plant to be a 'decentralized entity with no front or back, no left or right, but with a top or bottom' (Pouteau 2019, 89), with this last pair itself being a response to the Earth-Sun axis, at least in plants, and maybe even in humans. While these differences may make it difficult for humans to use their faculties to take the position of plants, there is arguably a frisson of anticipatory pleasure at the thought of being plant. A relief, possibly, at the thought of being decentralised with fewer binaries.

As Houle (2019) points out, for humans, and other animals, the process of gastrulation, an invagination of the embryo, creates a physical, material inner space, an internalisation of what was facing outwards, where, from mouth to anus is a turning in of what was originally an exteriority. As such, it is the material process of gastrulation that gives rise to a dualistic state of being, where the inside/self is prioritised over the outside/other. It is interesting that the human response to a crisis is, more-often-than-not, further separation, splitting off, coined as 'self-isolation' in the viral nomenclature of our times. We play to our material tendencies, you might say. While the result of gastrulation and subsequent separation can be imagined as a tube (Pouteau 2014), a doughnut (with a hole but no jam, of course) is possibly more suitable, not just because of its edibility, but because it is more clearly seen that the inside is in contact with what is outside and that there is a large surface area for doing so. The importance of this for knowing plants will be seen later. Pouteau (2014) argues that plants, on the other hand, go through no such process of gastrulation and invagination, and thus have neither an inside nor an outside, are 'noncentred' or 'unsplit' beings, living unlimited 'non-topos' existences. Put another way, they have no 'psychic interiority' (Marder 2012a). Pouteau (2014) regards this splitness of animals as being what separates them ontologically from plants, with the unsplit or open plant resonating with Deleuze and Guattari's (2005) notion of the rhizome in the sense that these creeping rootstocks signify the networked power to connect and proliferate. Pouteau also points out that plants never arrive at a 'completed state' as they lack the limits that human animals have, with this state of 'becoming' and 'unceasing synthesis' creating 'inlets and conveyances' with other critters. Importantly, Pouteau (2014) suggests that, due to their unsplit ontology, 'plants are incommensurable with animals' (3), including human animals, who have to 'enrapture, capture and consume the outside as food, intimacy and social life' (20). This chapter argues, however, that it is precisely the taking inside of plants in general, and medicinal plants in particular, which allows human bodies to work towards a commensurability that can benefit humans, sessile beings and wider relationships.

Pouteau (2014, 2019) explores 'open' as a way of thinking about plants, arguing that the openness of plants arises from their unsplit states, something which is evidenced by their unlimited material ways of being unfinished.

> Plants are open beings [because] their centre must be located in the open, non-Euclidean space and not 'inside' a putative 3D structure – be it a cell, a bud or a root-stock.
>
> (Pouteau 2019, 86)

Being unsplit or open is significant because it makes it difficult to discern parts from the whole and makes it even trickier to be clear about what is and is not 'plant'. This is a rather wonderful thought; plants have no centre, at least not in a way that we recognise the word, with no area that needs protecting at all costs. The idea of open, unsplit, flourishing, uncentred plant-ness is further explored by Marder (2013a, 115–30), through consideration of 'the plant that is not one', which means that 'above all, ...[a plant] is not one with itself', it has no self-identity, and does not reflect on self, that it does not enclose itself as animals do but that it has the 'potential for welcoming everything' – and that tap-rooting this potential is the job of vegetal democracy. This shared sufficiency is materially evident in how plants grow. A plant cutting can grow independently from the plant that it was severed from, suggesting that any self is likely to be dispersed and unrestricted, unlike a human sense of self, that is largely perceived as centred and confined (try pointing to yourself and see where your finger ends up). Not being one can also be seen in Coccia (2019), where plant life is conceived of as complete exposure, in absolute continuing and total communion with the environment, resulting in the impossibility of separating the plant in any way from the world it accommodates.

Plants coincide with the very substance of the world, so understanding the world cannot be separated from understanding plants. Non-separation from their environment suggests that plants cannot be reasoned to have a bounded inward vantage point. Marder (2012b, 26), puts it this way, noting that plants, unlike animals, are regarded as 'non-oppositional being(s)' to the point of 'merging with the milieu', which establishes plants as not being 'other' to their environment. Marder (2012c, 6) states that 'the plant does not organise its milieu', instead arguing that plants are made up of internal biochemical, hormonal, and synaptic communication networks as well as external communication routes that connect the plant to its environment. 'It is, thus, an open system, coupled with its environment ...' (Marder 2012c, 6). Seeds give a visible understanding of this plant non-separation: they germinate, growing shoots up and roots down, with no separation or traumatic break, as occurs in humans and other animals (Marder 2015a). Furthermore, as plants grow, the places where they are found change, with their own use of place depending on the positioning of new leaves, shoots, and roots (Marder 2012c). Referring to Hegel, Marder (2012b, 26) suggests that 'The plant is all about a visible extension without interiority'. These are enticing ideas for humans who arguably experience their sense of self as separated, restricted and centred.

Plantchantments

There are other matters of course, that give indications of enchanting differences between us humans and them plants. Marder (2013b) has elucidated distinctions by noting that change is the domain of plants not humans. Marder (2012c, 4) argues that '[a] rooted mode of being and thinking is characterised by extreme attention to the place and context of growth and, hence, sensitivity that at times exceeds that of animals'. Instead of moving and relocating to another place, or excavating and building around themselves, plants change through phenotypic plasticity and modular growth and thereby change the environment, from which they are inseparable. Plants change their state seasonally, with their visible busy-ness manifesting in front of our eyes, and reminding us to consider our own potential for change and growth. It is this sensitivity to their local ecology that they are never separate from and which is borne out of their rootedness, that suggests a fullness of attention is given by plants to where they are. Plants are mindful of where they are. Maybe *plantfulness* is a better conception, as it lacks the dualism that mind immediately conjures in us.

In addition, Marder (2015b) notes the modular development of plants, where leaf structures such as roots, branches, leaves and flowers are reproduced. This lack of vital organs means that plants can be alive and well even if parts are lost. Plant life is metaphysically superficial as it does not have a 'deep essence', a vital core that must be protected (Marder 2011). Superficiality is not a quality generally valued by humans, but seeing the benefits to plants could help us to consider the potential value of moving away from our interiors. Plants also have a fluctuating and indeterminate number of parts, with removed plant parts having the potential to become other plants (Marder 2014). Even, what could be thought of as the head can be lost and the plant remains vital – the head 'loses its transcendental privilege' (Marder 2011, 474). Marder (2013c, 124) refers to 'plant-thinking' as 'the non-cognitive, non-ideational, and non-imagistic mode of thinking proper to plants (hence, what I call "thinking without the head")'. While we self-ishly protect our brains, being with headless thinkers could open new ways of being in the world. Marder (2011, 485), referencing Deleuze and Guattari's 'A Thousand Plateaus' (2005), suggests that plants can be seen as 'a body without organs'. The possibility of seeing plants as life without vital organs is an attractive challenge, as a body without organs normally brings images of death.

Kalhoff (2019), like Pouteau (2019), sees value in attending to the idea of flourishing, suggesting that 'flourishing' has empirical and material meanings that are useful in comprehending plants. Flourishing indicates a 'good life'. Flourishing, as growth and flowering, implies that the plant is behaving as expected, manifesting life fully without impediment. 'Generous' could easily be added here to the characteristics of plants that enchant: the qualities of flourishing, merging with ecology, of being open, suggest a certain generosity. By having non-vital organs, plants can be generous to eaters, raising the question of whether we deserve them (Hedley, Shaw and Waddell 2023, forthcoming). Generosity also aligns with Attala's (2017) Edibility Approach that will be discussed later, where plants develop beneficial relationships with humans.

In summary, plants may be seen as open or unsplit beings, that are sensitive to their ecology and merge with it and shift shape in response to it, are without vital organs, think without a head and lead flourishing and generous lives, demonstrating alternative life ways that entangle with a porous world. They are the sort of critters that suggest possibilities of being different. For humanity, as split beings, with an inside and outside, these differences provide enchanting differences that takes us beyond ourselves. Let us now look at our similarities to plants. For this, we turn to post-millennial developments in plant science, to see something of ourselves in plants.

Sensing ourselves in plants: From palates to shyness and kin

The embryology of plant science itself reveals that considerations of human-plant similarities were not far away from the origins of the discipline. In a talk attended by the botanist Leonard Bastin, Darwin is reported as saying 'We must believe that in plants there exists a faint copy of what we know as consciousness in ourselves' (Bastin 1908, 551). Bastin (1908, 555) goes on to say that:

> In origin the animal and vegetable worlds appear to be indivisible, as though we may not dare to say that the plant is an intelligent being. There seems to be a field for a great deal of research, the

opening up of which will form a new and fascinating branch of botanical study.

And, of course, Darwin (1880, 573) wrote that:

> It is hardly an exaggeration to say that the tip of the radicle ... having the power of directing the movements of the adjoining parts, acts like the brain of one of the lower animals; the brain being seated within the anterior end of the body, receiving impressions from the sense organs and directing the several movements.

However, it would take over a hundred years before Bastin's anticipatory excitement and Darwin's radical radicle suggestion were to become rooted in ground fertile enough to produce new growth.

In 1984, the Nobel Prize winner for physiology and medicine, Barbara McClintock, in her acceptance speech, referring to plants, said that 'A goal for the future would be to determine the extent of knowledge the cell has of itself and how it utilises this knowledge in a thoughtful manner when challenged' (Trewavas 2015, v). Her biographer quotes McClintock as saying, in reference to her research into corn:

> I start with the seedling, and I don't want to leave it. I don't feel I really know the story if I don't watch the plant all the way along. So I know every plant in the field. I know them intimately, and I find it a great pleasure to know them.
> (Keller 1983, 198)

The value of spending time with plants in order to have direct knowledge of them can also be seen in Gagliano and Grimonprez's (2015, 150) 'ecologically driven approaches, where the "cultural background" of plants is taken into account'. To have any hope of understanding culture, as any ethnographer knows, immersion in local ecologies is required, and that takes time.

Moreover, Gagliano, Ryan and Vieira (2017, xiii) argue that recent developments in plant science research 'involve[s] a real celebration of "plantness"'. This is evident in the work of high-profile international researchers, who, while dispersed, are forming a web that is sending out ripples within and beyond the academy. What had seemed like unchallengeable philosophical assertions of the ontological difference and superiority of animal life over plant life are being questioned by studies that demonstrate plants as manifesting intelligent behaviour seen through their sensory awareness, processing of information, ability to anticipate, remember, learn and make decisions (see: Brenner et al. 2006; Chamovitz 2013; Dudley and File 2007; Garzon and Deijzer 2011; Mancuso and Viola 2015; Pollan 2013; Simard 2009, 2012; Simard et al. 2012; Trewavas 2015). These new directions in plant science are one way in which we can see ourselves in plants. Not surprisingly perhaps, this is largely through work that investigates plant senses, behaviour, communication and memory. The work of some of these central researchers highlight the question of similarities between humans and plants, as the following illustrates.

> the tree will remember it was touched. But it won't remember you.
> (Chamovitz 2013, 175)

Several scholars explore plants' sensitivities to light, touch and smell. Chamovitz (2013), for example, describes how plants can distinguish between different colours and can tell if the shirt you are wearing is blue or a red, although not that it is a shirt. Similarly, plants are aware of the developing sunset. They also know which way is up or down, with this being achieved via specialised cells that have a similar function to human ears that position the plant within the wider environment. In addition, plants are acutely aware of aromas, for example, producing chemical volatiles if attacked by predators which induces nearby plants to build up chemicals in their leaves that are toxic to these predators, or which attract the predators of these predators. The enemy of my enemy is my friend.

Mancuso and Viola (2015) argue plants can be seen to have 'palates', with taste through registration of the soil contents being a process that can identify mineral salts in minute concentrations, and focus root growth towards the source, foraging for this nourishment. The importance of touch or, even more impressive perhaps, the ability to avoid it, can be seen in the 'crown shyness' (Mancuso and Viola 2015) of some trees, where they avoid touching each other's crowns when growing in close proximity, suggesting some sort of agreement to respect each other's space.

Mancuso and Viola tell the story of the famous scientist Lamarck (1744–1829) who asked his colleague to push some *Mimosa pudica* specimens, plants that are known to retract their leaves when touched, on a cart through the uneven streets of Paris. He reported back that initially the plants closed their leaves in response to the vibrations of the cart but soon reopened them when the experience proved not to be dangerous. Interestingly, Gagliano et al. (2014) performed a similar experiment but established new findings. After repeatedly dropping *Mimosa pudica*, retractions stopped as expected. However, Gagliano et al. also found that after a month with no further droppings the plants continued to not retract their leaves after being dropped, suggesting that a cellular calcium signalling network may be responsible for the apparent memory accessed by the plant.

Gagliano (Gagliano 2012; Gagliano et al. 2017) is at the root tip of explorations of the emergent field of plant bioacoustics, showing that plants produce sounds that are not accounted for by cell elongation or water transport. Unsure of the reason, consideration is given to the possibility that these emissions warn neighbours of impending herbivory thereby influencing insect behaviour. If bacteria can communicate using ultrasonic waves, Gagliano asks whether plants might do the same. Gagliano et al. (2017) have shown that plant roots move towards the sound of water in pipes, i.e. without accessing moisture gradients, but that, given the choice their roots move towards water gradients rather than water enclosed by pipes, demonstrating the selection of the most advantageous cue for quenching thirst.

Karban (2017) notes that plants are able to tell the difference between shade produced by other plants and that from objects that do not move, and will choose to grow away from their competitors for light. Karban (2015) also describes the sensitivity of climbing tendrils to touch as being greater than that of humans and suggests that plants may have as yet undescribed senses.

Dudley and File (2007) explore how beach weed sea rocket (*Cakile edentula*) uses root interactions to identify whether it is growing with siblings or unrelated plants of the same species. If growing with kin, these plants are less competitive for resources and use existing supplies for aerial growth and hence kin reproduction. If grown with stranger plants, however, allocation of resources to roots for competitive foraging is increased. Talking of kin, Gagliano and Grimonprez (2015, 149) suggest that the chemical messages that volatile emissions carry are more effective when received by kin than by strangers and, importantly, that related plants may be identified by 'leaf gestures', suggesting a process akin to 'cultural transmission'.

Simard (2009, 2012; Simard et al. 2012) has found that trees in Douglas fir forests share resources amongst themselves using underground mycorrhizal fungal networks that connect trees' root systems and share carbon-based food with young fir seedlings. The oldest 'mother trees' are the most active 'hubs', sharing resources with shaded seedlings while still not tall enough to access sufficient light. They also lend sugars to beech trees when in surplus, with these resources being returned at a later date. These exchanges promote a healthier forest and, importantly for plants and humans, a large volume of photosynthesis. Fir trade, you might call it. Needless to say, or maybe that should be needles to say, as we are talking of fir trees, this has been referred to as the 'wood wide web'.

This brings us to the last, and probably most obvious, demonstration that plants are more like animals than had previously been thought. Taking inspiration from Pavlov's dogs experimentation into associative learning, Gagliano et al. (2016) used pea plants (*Pisum sativum*), instead of dogs, with an air flow via a fan and light as the reward, instead of the sound of a bell and food. The plants conditioned by the air flow grew towards the fan even without the presence of light, appearing to demonstrate associative learning in plants. Plants, it seems, can identify and anticipate reward.

Anthropomorphism versus anthropocentrism

The examples above demonstrate plantabilities that resonate with us two-legs. Plants can distinguish between different coloured lights, identify aromas, call the enemies of enemies to intervene, forage for nutrients and water, identify what is dangerous and what is not, possess sensitivity to touch (the ability to refrain from touching the crowns of others), an awareness of what is producing shade, can identify kin through interactions and gestures (and share with this kin), exchange with others, and anticipate reward. It is interesting that it is scientists that have arguably anthropomorphised plants, through the questions they ask, their experimental designs and descriptions of results. Bennett (2010) has come out to bat on behalf of the value of cultivating a degree of anthropomorphism, the attribution of human characteristics to nonhumans, in order to oppose the self-absorption of humans. Bennett argues that anthropomorphising is not a terminal activity, rather, that viewed through a vital materialist lens:

> An anthropomorphic element in perception can uncover a whole world of resonances and resemblances – sounds and sights that echo and bounce far more than would be possible were the universe to have a hierarchical structure.
>
> (Bennett 2010, 99)

While Bennett points out the risks that travel with anthropomorphising, including romanticism and the divinisation of nature, she argues that anthropomorphising pushes against anthropocentrism. Resonances between human persons and more-than-humanity mean that separation is less likely and entangled conversations more likely. This understanding of anthropomorphism, as working against anthropocentrism by uncovering resonances with the more-than-human, can be seen in Harvey's 'new' animism (Harvey 2005, 2015), and also in Hall's *Plants as Persons* (Hall 2011) where the world is made up of persons, only some of whom are humans, even if the 'as' in the title recognises the difficulties faced in such a (re)conceptualisation.

While seeing ourselves in plants through the traction of plant similarity, as articulated by recent plant science, combined with the enchantment of plant difference thanks to vegetal philosophy, can help us to entertain different ways of being, as split beings we *need* to take plants inside us. To this end we turn, via Attala's Edibility Approach (2017), to plant ecology as medicine for humans in order to look at what plants do both in their ecologies and inside us.

Edibility, medibility, secondary plant metabolites and ecology as medicine

The 'phyto-centric framing', that is central to Attala's (2017) Edibility Approach, pays attention to the ways that plants impact human bodies via their edibility, suggesting that plants are, in fact, 'inciting behaviours' through the conversation of digestion. Drawing on recent work in more than human, multi-species and new materialities thinking, digestion is described as relational, with plants influencing whoever digests them in ways that benefits plants. Attala also argues that the relationality of the Edibility Approach is ecological in that life is always co-produced. While pharmacology defines pharmacodynamics as what a drug (including plant medicines) does to a body, an Edibility Approach, on the other hand, looks at what plants do to people, including what plants get people to do, which we could call phytodynamics, and is played out over a much longer time period. Attala places the Edibility Approach within a new materialities turn in the academy, arguably driven by anthropology, where materials are 'lively subjects of study', where matter is found in theories and where the world is 'produced entirely of, with and from matter.' This matter is conceived of as 'leaky' and 'porous', with the properties of any particular matter being the limiting factor in their movement across boundaries. 'Consequently, each relationship is stipulated and prescribed by the brute physico-chemical parameters of that engagement' (Attala 2017, 130). The Edibility Approach 'all but embraces the chemistry of interactivity' (Attala 2017, 130). We will also head towards chemistry, but seek to head it off at the pass before getting seduced by its apparent precision and certainty.

I would like to both compress and stretch the Edibility Approach to suggest a medibility approach that includes medicinality as a subset of edibility in plant-human relationships. The notion of stretching attends to the ecological extension of these intimate relationships.

As split beings that look outwards, one could argue that much of human knowledge comes from taking things in. If the ability to know plants comes from taking them in, the ingestion of medicinal plants provides a convenient direction to explore. There are multiple different cultural traditions and recipes associated with ingestion. Here I am going to limit myself to understandings rooted in the diverse practices of Western herbal medicine (Waddell 2020), to demonstrate how ingestion offers a plurality of possibilities, including plant chemicals, qualities, virtues, energetics, properties and actions which can bind, inhibit, stimulate, astringe, heat, soften, moisten, restore, cool, balance, move, regulate, open, detoxify, calm, lift and repair, to name but a few possibilities (Bone and Mills 2013; Holmes 2020; Tobyn et al.

2016; Wood 2004). Approaching just one of these threads, namely phytochemistry, not only offers an understanding of the specific biological and material processes of how plants have medicinal effects on humans, it also, perhaps surprisingly, points to fruitful ways of understanding what it means to be medicinal in broad terms. To this end, this section explores the relationality of plant medicine through chemical commonalities found in the ecologies of plants and their medicinal properties.

When it comes to plants, the matter of medicinal plants is also the matter of ecological participation. To explore this, let us first turn to what is regarded as a foundational binary of plant chemistry. Plant cells produce 'primary metabolites' that are involved in growth and metabolism, such as carbohydrates, vitamins, organic acids and amino acids. However, they have also evolved biochemical pathways that manufacture a range of chemicals called plant secondary metabolites that do not play a role in primary metabolic processes, and which are not necessary for the day-to-day survival, growth, development and reproduction of plants, even if they do increase overall fitness in several ways. Interestingly, there are about 200,000 secondary metabolites across all plant species, while there are only about 10,000 primary metabolites (Kennedy 2014). This difference is due to the origin of primary metabolites in the common single celled ancestors of all plants, while secondary metabolites developed along the speciation of plants in response to their local environments relatively recently. This suggests that the high number of secondary metabolites reflect the diversity and complexity of ecologies. These secondary metabolites are generally regarded as having a number of evolutionary roles (Kennedy 2014), specifically attraction and defence. There is, for example, the attraction of pollinators via colour and volatile emissions, the predators of herbivores by volatile emissions, and ant guards by paying them extra floral nectar. On the other hand, the defensive role of secondary plant metabolites can be seen in their antimicrobial activities, in their toxicity to herbivores, in their deterrence of egg laying, and in their impact on nectar components in order to deter nectar robbers.

However, this simplistic picture sidesteps a phytochemical consideration or explanation of the types of behaviours, communications, agency or ontology that recent plant science and vegetal philosophy point towards. In fact, it is only recently that hypotheses that allow chemodiversity to be conceptualised as a driver in ecological processes have been released (Kessler and Kalske 2018). This brings an opportunity to allow secondary metabolite hypotheses to potentially find their way to interface with vegetal philosophy and botany. Because primary metabolites have a well-described and life-sustaining yet limited role in plant life, it seems reasonable to suggest that, at a chemical level, it is the rich realm of secondary plant metabolites that allow and facilitate, or even drive, depending on the degree of agency that we give to phytochemical matter, the revolutionary ideas emerging from vegetal philosophy and recent plant sciences, as described above. Thus, it appears likely that secondary metabolites are associated with the openness and unsplitness of plants – with their acute positional awareness, their porosity with their surroundings (if indeed it is possible for those who are not separate to have 'surroundings' at all), their ability to respond to local conditions, to comprehend without an identifiable organ of comprehension, with their generosity and their lack of the indispensable. Plant secondary metabolites are also likely to be inseparable from a range of sense faculties, including to touch, colour, and sound. Phytochemically, it appears that secondary metabolites are also responsible for the medicinal properties of plants (Bone and Mills 2013; Heinrich et al. 2018; Williamson 2003). Possible mechanisms include covalent modifications of proteins and DNA bases, interactions with biomembranes and antioxidant properties (Wink 2015), even if traditional and other texts have their own range of explanations for what medicinal plants can do and are more clinically useful.

From the above, it follows that when a plant is medicinal, it is due to the plant's ecological relationship with the world. Medicinal plants, when ingested, engage with 'us' as the world through secondary metabolites. This can be seen as an example of 'worlding' that Donna Haraway (2016) refers to. From a materio-existential perspective, we are no longer separate. As plants are porously contemporaneous with the environment, human engagement with plants extends that relationship towards the landscape ecology, world. Whole and healed, you might say. No longer being separate and being well, may be one and the same thing. Maybe the simple truth is that the benefits of plant medicines are arrived at through us becoming less separate, even non-separate, from our ecology, which is now no longer inside or outside of us, but of which we are part.

Seeing plants as being medicinal to humans through their ecologies has implications for human relationships with the more-than-human world in that it is in the interest of human animals to look after plants, not just as bounded individuals or species, which they have been perhaps erroneously presented as, but as individuals and species that are not separate from wider ecologies.

Taking inspiration from Attala's Edibility Approach (2017), a medibility angle suggests that some plants possess the gravity of attraction and through ingestion (or topical application) are able to penetrate bodies, tangibly reducing the illusion of separation that our material existence seems to present as common sense. While it is possible, as suggested above, to conceive of herbal medicine as beneficial because of plants' wider ecological relationships, it is ultimately lives lived consciously *with* plants and therefore the world, that provides a new understanding of, and route to, a sustainable approach to health through and with plants.

A taste of medicine as ecology

> If a plant jumps out at you, you must make a tea.
> (Hedley, in conversation)

The state of secondary plant metabolite chemistry is not nearly developed enough to fully understand how these chemicals clinically materialise health in humans. Despite only being in its infancy in developing knowledge of how medicinal plants live in their ecologies, the matter of phytochemicals offers us an opportunity via attention to ingestion. As split beings, human animals, just like other animals, must take medicinal plant matter inside. The gastro-intestinal tract provides a surface area of up to 300 square metres for conversations with ingested medicinal plant matter.

It is possible to acquaint yourself with plant chemistry through tasting and ingesting (Hedley, Shaw and Waddell 2023, forthcoming). Sourness can indicate, for example, organic acids, flavonoids, anthocyanidins and tannins, while bitterness can detect bitter alkaloids, iridoids, anthraquinones, cardioactive glycosides, salicylates and sesquiterpenes. A sweet taste may signal the presence of polysaccharides, saponins and isoflavones and a pungent taste, the presence of pungent alkaloids, essential oils, phenolic acids, resins, coumarins and glucosinolates, while a salty taste can suggest salts and minerals. Many of these are likely to be active, in yet unknown ways, in plant-ecology relationships.

Plant tea tasting facilitates human access to worlding by plants, mediated by secondary plant metabolites. A tea tasting can take you all over the world. While taste can be used to help identify constituents, it is arguably more useful to say that it is a method of directly knowing the plant, which includes its phytochemical constituents of course. Tasting provides an understanding of qualities, and what the plant does in and with the matter of the body. Plant flourishing and generosity allows this, in fact, encourages or even demands it in a Medibilities approach. For Hedley and Shaw, the sense of taste, with all the pathways and grooves from which it is formed, is awoken most easily through making and taking of a tea of the medicinal plant. This is a fundamental tool for practice, which requires knowing plants intimately.

In all traditional medicine systems, the properties and qualities of a medicine can be revealed through tasting it. Hedley and Shaw note that methods of tasting 'means the appreciation of a herb using all the senses ...' (Hedley and Shaw 2019, xxxvii) and includes the traditional European concept of 'appropriations', that is:

> the appreciation of a remedy by the whole body not just by the special senses. The taste in your mouth is the beginning of your understanding of a remedy but a full appreciation comes only after feeling how it interacts with your body and by observing its actions in a range of people.
>
> (Hedley and Shaw 2019, xxxvii–xxxviii)

They describe a simple method for tasting, which deliberately allows for complexity to come through, where weak teas and the avoidance of technical language are advised. It is a twelve step process involving noticing the colour, aroma, sipping and swishing, noticing texture and sensation, looking for the classic tastes and textures of bitter, sour, sweet, salty, savoury, pungent, dry and moist; further sipping and drinking in, watching the herb as it travels into your body, watching for appropriations i.e. the parts of your body (tissues, organs, limbs, areas, etc.) where the activity of the herb is mainly felt; paying attention to what it is doing, how it is moving – up or down, inward or outward, its pace of movement, whether it expresses itself directly or in waves, whether it has a warming or cooling effects; if and how it affects mood, for example, inducing peace, or edginess and stimulation, before considering what the herb 'may be like', possibly 'a season, a time of day, a person you know…' (Hedley and Shaw 2019, xxxviii) and ending with a consideration of what sort of person may benefit from this herb. This method elucidates the body's porosity and its receptivity to the agency of the properties of ingested plant medicine.

This appreciation of the qualities of herbs through tea tasting can usefully be considered by taking a sip of Ingold's work on correspondence (Ingold 2016). Ingold suggests that correspondence 'is the process by which things literally answer to one another over time' (2016, 14), with this conversation growing from the related principles of 'habit', 'agencing' and 'attentionality'. While habit refers to the processes by which we are always shaping the conditions in which we live and will live, Ingold describes agencing as 'doing undergoing', suggesting that it is action itself that produces agency, rather than agency being 'given in advance of action' (Ingold 2016, 17). The third principle of correspondence is attentionality, which Ingold describes as 'going along with things, opening up to them and doing their bidding' (Ingold 2016, 19), noting that this *with*-ness 'saves the other from objectification by bringing it alongside as an accomplice' (Ingold 2016, 19).

Herbal tea tasting, in its whole-body appreciation of herbs, can be seen as the answering of herbs and human people 'to one another'. This conversation arises from 'doing undergoing' through smelling, sipping, swallowing and paying attention. Importantly, the agency that is produced out of this process requires that the taster must 'go along with' the herbs and 'do their bidding'. A sort of surrender to the 'accomplice' that is the herb is required of the person, with this involving going beyond thinking. In herbal tea tasting, an impression is made on an organ or tissue, and a conversation follows, with this organoleptic experience allowing the herb to be followed and appreciated.

In the following passage, Hedley and Shaw (2014, 1–3) give an account of the agencing that can result from the qualities, including chemistries, of herbs:

Heat cherishes the vital spirits and is warming, softening, diffusive, active, ambitious and extrovert. It thins thickened humours, takes away weariness, abates inflammation (which is after all a form of congestion), relieves pain due to congestion or spasm, moves and thins the blood ... Cold is firming, holding, consolidating, conserving, slowing, calming, centred and introvert. It modifies excess heat, qualifies the heat of food (so that it can be better digested), cools the blood, refreshes the spirits when they are suffocated, brings down inflammatory swellings and stops the pores ... Traditionally bitters are cooling and downward moving. The downward movement can be felt clearly with straight up bitters such as Gentian and it is this movement that gives us the key to the general properties of bitters – cooling (bringing heat down) and digestive (moving digestion along). Bitters also straighten and hold the centre ('consolidating' in TCM) hence the popularity of bitter tonics ... GENTIAN is almost a pure iridoid bitter and the best for tasting purposes. It appropriates to the spleen which is responsible for turning what you eat into you ... Anthraquinones have an earthy – bitter taste which gives them a strong downward movement, useful in clearing the bowels ... Sesquiterpenes are closely related to volatile oils and often found with them. They are bitter and cooling but also pungent and warming... Saponins taste sweet and strengthening ... In traditional medicine the sweet taste builds up (although in excess it will weaken) ... Polysaccharides are also sweet tasting – although they may need to be held in the mouth a while in order to appreciate their full sweetness. The combination of polysaccharides and saponins makes for some excellent strengthening remedies such as GINSENG ... Volatile oils (have a) common theme of a pungent, warming taste which rises, disperses and relaxes ... Resins are pungent and warming but also very sticky and bitter ... the sticky quality of (resinous) MYRRH makes it especially useful for healing in wet conditions, such as in the mouth and for weeping wounds.

Medicinal plants act in human bodies. They diffuse, remove, abate, relieve, move, thin, firm, hold, conserve, slow, modify, bring down, stop and strengthen lively matter through porous processes and semipermeable membranes. Plants appropriate *to* particular tissues, organs or other materials of the body co-constructed within the shared corporeal environment. We are now a plant's ecology, and rejoined or re-membered with the world external to flesh. Such entanglements of humans with plants are ecological actions that present human splitness, separation and exceptionalism in a new light.

Michel Serres (2016) refers to the senses as being tangled together to form knots, with the body constructing itself through its sense organs. Serres maintains that:

> Knowledge cannot come to those who have neither tasted nor smelled. Speaking is not sapience…[because] sensation, it used to be said, inaugurates intelligence. Here, more locally, taste institutes sapience.
>
> (Serres 2016, 154–5)

This is true even if the capacity of taste for rich experience often overwhelms words. The body remains 'the medium of intuition, memory, knowing, working and above all, invention.' (Serres 2012, 34).

From this perspective, medicinal tea tasting allows the knotted senses to infuse together, with the experience reconstructing the body through Ingold's 'going along with', 'opening up' and 'doing the bidding' of medicinal plants. These steps, of course, require that we are genuinely curious about, and willing to hang out with sessile green critters, including as medicinal persons, and that we can accept the agencing of plants as we interact with them – activities, the absence of which, have arguably led in part to the Anthropocene. Tasting and knowing plants as medicinal through their ecology is one thing. If we add to this being enchanted by the unsplitness and non-separation of plants as explored in vegetal philosophy, while also recognising ourselves in plant behaviours as suggested by recent plant science, we will be more capable of forming human-plant ecologies with which we and all sorts of others can live.

References

Attala, L., 2017. '"The Edibility Approach": Using edibility to explore relationships, plant agency and the porosity of species' boundaries', *Advances in Anthropology* 7, 125–45. doi:10.4236/aa.2017.73009.

Bastin, L., 1908. 'The intelligence of the plant', *Pall Mall Magazine* 42(187), 550–8.
Bennett, J., 2010. *Vibrant Matter: A Political Ecology of Things*. Durham, NC and London: Duke University Press.
Bone, K. and Mills, S., 2013. *Principles and Practice of Phytotherapy: Modern Herbal Medicine*. London: Elsevier.
Brenner, E. D., Stahlberg, R., Mancuso, S., Vivanco, J., and Van Volkenburgh, E., 2006. 'Plant neurobiology: An integrated view of plant signaling', *Trends in Plant Science*, 11(8), 413–19.
Chamovitz, D., 2013. *What a Plant Knows*. Oxford: One World.
Coccia, E., 2019. *The Life of Plants: A Metaphysics of Mixture*. Cambridge: Polity Press.
Darwin, C., 1880. *The Power of Movement in Plants*. London: John Murray.
Deleuze, G. and Guattari, F., 2005. *A Thousand Plateaus: Capitalism and Schizophrenia*. Minneapolis, MN: University of Minnesota Press.
Dudley, S. and File, A., 2007. 'Kin recognition in an annual plant', *Biology Letters* 3, 435–8.
Garzon, P. and Deijzer, F., 2011. 'Plants: adaptive behavior, root-brains, and minimal cognition', *Adaptive Behavior* 19(3), 155–71.
Gagliano, M., 2012. 'Green symphonies: a call for studies on acoustic communication in plants', *Behavioral Ecology* 24, 789–96.
Gagliano, M. and Grimonprez, M., 2015. 'Breaking the silence – language and the making of meaning in plants', *Ecopsychology* 7(3), 145–51.
Gagliano, M., Renton, M., Depczynksi, M. and Mancuso, S., 2014. 'Experience teaches plants to learn faster and forget slower in environments where it matters', *Oecologia*. 175(1), 63–72.
Gagliano, M., Vyazovskiy, V., Borbely, A., Grimonprez, M. and Depczynski, M., 2016. Learning by Association in Plants. *Scientific Reports*, 6: 38427.
Gagliano, M., Grimonprez, M., Depczynski, M. and Renton, M., 2017. 'Tuned in: Plant roots use sound to locate water', *Oecologia* 184(1), 15–60.
Gagliano, M., Ryan, J. C. and Vieira, P., 2017. *The Language of Plants: Science, Philosophy and Literature*. Minneapolis, MN: University of Minnesota Press.

Hall, M., 2011. *Plants as Persons*. Albany, NY: State University of New York.

Haraway, D., 2016. *Staying with the Trouble: Making Kin in the Chthulucene*. Durham, NC and London: Duke University Press.

Harvey, G., 2005. *Animism: Respecting the Living World*. London: C. Hurst and Co.

Harvey, G., 2015. *The Handbook of Contemporary Animism*. New York, NY: Routledge.

Hedley, C. and Shaw, N., 2019. *A Herbal Book of Making and Taking*. London: Aeon Books.

Hedley, C. and Shaw, N., 2014. *Galenic Energetics Applied to Medicinal Plant Constituents*, pp. 1–3. Unpublished research.

Hedley, C., Shaw, N. and Waddell, G. (eds), 2023. *Plant Medicine: A Collected Teachings of herbalists Christopher Hedley and Non Shaw*. London: Aeon Books. Forthcoming.

Heinrich, M., Barnes, J., Prieto-Garcia, J., Gibbons, S. and Williamson, E., 2018. *Fundamentals of Pharmacognosy and Phytotherapy*. London: Elsevier.

Holmes, P., 2020. *The Energetics of Western Herbs: A Materia Medica Integrating Western and Chinese Herbal Therapeutics*. Fourth Edition. London: Aeon Books.

Houle, K., 2019. 'Facing only outwards? Plant bodily morphogenesis and ethical conceptual issues', in A. Kallhoff, M. Di Paola and M. Schorgenhumer (eds), *Plants and Ethics: Concepts and Applications*. Abingdon and New York, NY: Routledge, pp. 70–81.

Ingold, T., 2016. 'On human correspondence', *Journal of the Royal Anthropological Institute* 23, 9–27.

Kalhoff, A., 2019. 'The flourishing of plants: a neo-Aristotelian approach to plant ethics', in A. Kallhoff, M. Di Paola and M. Schorgenhumer (eds), *Plants and Ethics: Concepts and Applications*. Abingdon and New York, NY: Routledge, pp. 51–8.

Karban, R., 2017. 'The language of plant communication (and how it compares to animal communication)', in M. Gagliano, J., Ryan and P. Vieira (eds), *The Language of Plants: Science, Philosophy, Literature*. Minneapolis, MN: University of Minnesota Press, pp. 3–26.

Karban, R., 2015. *Plant Sensing and Communication*. Chicago, IL: University of Chicago Press.

Keller, E., 1983. *A Feeling for the Organism: The Life and Work of Barbara McClintock*. New York, NY: W. H. Freeman.

Kennedy, D., 2014. *Plants and the Human Brain*. New York, NY: Oxford University Press.

Kessler, A. and Kalske, A., 2018. 'Plant secondary metabolite diversity and species interactions', *Annual Review of Ecology, Evolution and Systematics* 49, 115–38.

Mancuso, S. and Viola, A., 2015. *Brilliant Green, the Surprising History and Science of Plant Intelligence*. Washington, DC: Island Press.

Marder, M., 2011. 'Vegetal anti-metaphysics: Learning from plants', *Continental Philosophy Review* 44, 469–89.

Marder, M., 2012a. 'The life of plants and the limits of empathy', *Dialogue* 51(2), 259–73.

Marder, M., 2012b. 'Resist like a plant! On the vegetal life of political movements', *Peace Studies Journal* 5(1), 24–32.

Marder, M., 2012c. 'Plant intentionality and the phenomenological framework of plant intelligence', *Plant Signaling and Behavior* 7(11), 1–8.

Marder, M., 2013a. 'Vegetal democracy: The plant that is not one', in A. Magun (ed.), *Politics of the One*. London: Bloomsbury Academic, pp. 115–30.

Marder, M., 2013b. 'Of plants, and other secrets', *Societies* 3, 16–23.

Marder, M., 2013c. 'What is plant-thinking?' *Klesis–Revue Philosophique* 25, 124–43.

Marder, M., 2014. 'For a phytocentrism to come', *Environmental Philosophy* 11(2), 237–52, doi: 10.5840/envirophil20145110.

Marder, M., 2015a. 'The sense of seeds, or seminal events', *Environmental Philosophy* 12(1), 87–98, doi: 10.5840/envirophil201542920.

Marder, M., 2015b. 'The place of plants: Spaciality, movement, growth', *Performance Philosophy* 1, 185–94.

Mentz, S., 2019. *Break Up the Anthropocene*. Minneapolis, MN: University of Minnesota Press.

Pillitteri, L. J., Guo, X. and Dong, J., 2016. 'Asymmetric cell division in plants: mechanisms of symmetry breaking and cell fate determination', *Cell Mol Life Sci.* 73(22), 4213–29, doi: 10.1007/s00018-016-2290-2.

Pollan, M., 2013. 'The intelligent plant: Scientists debate a new way of understanding flora', *The New Yorker*, 23 and 30 December 2013. [online] Available at: *http://www.newyorker.com*. Accessed June 2021.

Pouteau, S., 2014. 'Beyond "second animals": Making sense of plant ethics', *Journal of Agricultural and Environmental Ethics* 27, 1–25, doi: *10.1007/s10806-013-9439-x*.

Pouteau, S. 2019. 'Plants as open beings: From aesthetics to plant-human ethics', in A. Kallhoff, M. Di Paola and M. Schorgenhumer (eds), *Plants and Ethics: Concepts and Applications*. Abingdon and New York, NY: Routledge, pp. 82–97.

Rasmussen, C. G. Humphries, J. A. and Smith, L. G., 2011. 'Determination of symmetric and asymmetric division planes in plant cells', *Annual Review of Plant Biology* 62, 387–409, doi: *10.1146/annurev-arplant-042110-103802*.

Schwagerl, C., 2014. *The Anthropocene: The Human Era and How It Shapes Our Planet*. Santa Fe, NM: Synergetic Press.

Serres, M., 2012. *Variations on the Body*. Minneapolis: University of Minnesota Press.

Serres, M., 2016. *The Five Senses*. London: Bloomsbury Academic.

Simard, S. W., 2009. 'The foundational role of mycorrhizal networks in self-organization of interior Douglas-fir forests', *Forest Ecology and Management*, 258S, 95–107.

Simard, S. W., 2012. 'Mycorrhizal networks and seedling establishment in Douglas-fir forests', in D. Southworth (ed.), *Biocomplexity of Plant–Fungal Interactions, First Edition*. London: John Wiley and Sons, pp. 85–107.

Simard, S. W., Beiler, K. J., Bingham, M. A., Deslippe, J. R., Philip, L. J. and Teste, F., 2012. 'Mycorrhizal networks: mechanisms, ecology and modelling. Invited review', *Fungal Biology Reviews* 26, 39–60.

Subcommission on Quaternary Stratigraphy. [online] Available at: *http://quaternary.stratigraphy.org/working-groups/anthropocene/*. Accessed May 2021.

Tobyn, G., Denham, A. and Whitelegg, M., 2016. *The Western Herbal Tradition: 2000 years of Medicinal Plant Knowledge*. London: Singing Dragon.

Trewavas, A., 2015. *Plant Intelligence and Behaviour*. Oxford: Oxford University Press.

Vernadsky, V., 1998. *The Biosphere: Complete Annotated Edition.* Göttingen: Copernicus.

Waddell, G., 2020. *The Enchantment of Western Herbal Medicine: Herbalists, Plants and Nonhuman Agency.* London: Aeon Books.

Weart, S. R., 2008. *Discovery of Global Warming.* Revised and expanded edition. Cambridge, MA and London: Harvard University Press.

Williamson, E., 2003. *Potters Herbal Cyclopedia.* Saffron Walden: The C. W. Daniel Company Limited.

Wink, M., 2015. 'Modes of action of herbal medicines and plant secondary metabolites', *Medicines* 2, 251–86, *doi: 10.3390/medicines 2030251.*

Wood, M., 2004. *The Practice of Traditional Western Herbalism: Basic Doctrine, Energetics and Classification.* Berkeley, CA: North Atlantic Books.

Zernicka-Goetz, M., 2020. *The Dance of Life: Symmetry, Cells and How We Become Human.* London: W. H. Allen.

7 ESCAPING TO THE GARDEN AND TASTING LIFE

Sarah Page

Introduction

> When we were not allowed to go anywhere else, getting out and tending the garden was an enormous help – and even when we were allowed out over the summer, the garden and its health always came first.
>
> (Sharon)

This chapter draws on an ethnographic and auto-ethnographic study of gardeners who were growing their own food whilst adhering to the 'stay-at-home' order in England, UK during the year 2020–2021 because of the Covid-19 pandemic. It argues that the act of gardening offered reassurance of life's continuities during the pandemic's upheaval and that being involved in growing one's own food provided a deeper sensorial appreciation of nourishment to those involved. The study used phenomenological accounts to explore the meaning of gardens and growing vegetables during lockdown. According to the media, many people across Britain turned to gardening at the beginning of the Covid-19 pandemic and sought an engagement with the outdoors as a means of coping with the restrictions imposed on social mobility and interactions. There is a plethora of research on benefits of gardening for health (Soga, Gaston and Yamaura 2017; Thompson 2018), and for community and social engagement (Harper and Afonso 2016; Wesselow and Mashele 2019). Whilst gardening as a hobby may be considered a current trend, and more so during Covid, it has been extolled in history as a practice necessary for human health and well-being in many cultures, and not just as a means of securing a food source. In the UK, the Allotment Movement and Garden Cities of the late 1800s were designed to encourage urban dwellers' contact with nature, because it was 'necessary for the complete physical and social development of humankind' (Willes 2014, 292).

162   PLANTS MATTER

Peggy·
22 Mar ·

I'm sharing these pictures, both taken last year. My favourite one of the garden was when my best friend came over during a lockdown ease. It makes me feel happy every time I see it. My 2nd job (a carer for ind. with MS) - bought me an engraved trug which I adore. The other pic. is my day job in ophthalmic surgery. We didn't have any furlough. Spending time with my plants and in the garden completely absorbs and fulfils me. I found it wonderfully. distracting. I'm about to start my 3rd year gardening. 😊 🌷 🌱

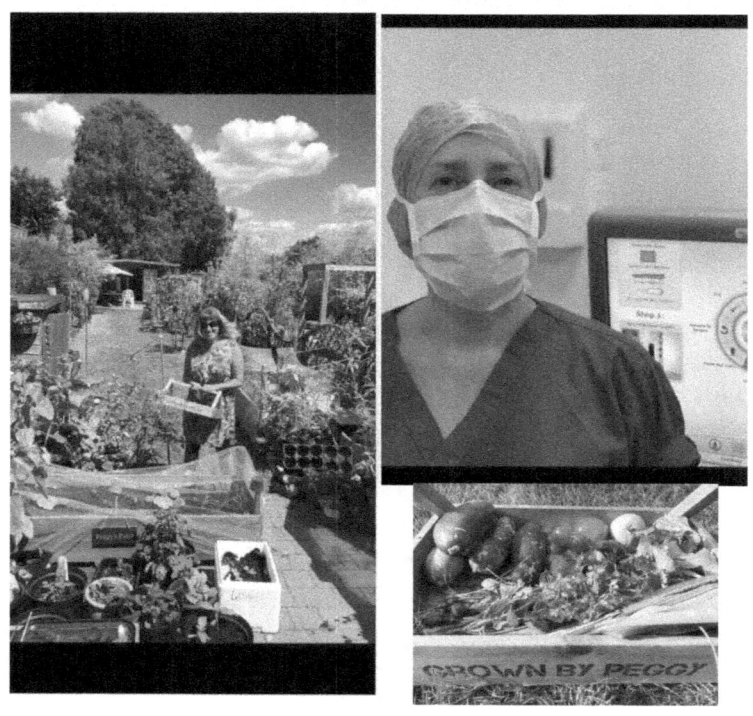

Fig. 7.1   'Spending time in the garden completely absorbs and fulfils me' (Peggy).

The term 'nature' is a thoroughly contested concept, and much has been written about the problems of using it (Descola 2013). It suggests a backdrop to human lives or a location that people can visit. Following Strang's (2006) comments about how waters' flows trouble one's ability to successfully locate nature as a state distinct from culture, gardens provide a similar problematising of concepts as they can be conceived as straddling the boundary between nature and culture. Neither fully natural nor simply cultural, gardens offer another example of how people and nature intermingle. My study demonstrates not only how people use their gardens but how deep-rooted relationships with plants develop over time and thus gardens become important places where, through the kind of sensory attention that pottering allows, lives grow and entangle (Ingold 2010). Gardeners' accounts demonstrate that relationships with the garden manifest as multiple complex ecological interactions between a variety of players including non-human animals, plant-life, diverse objects and materials, and events such as the weather and seasonal changes. Describing these relationships and identifying the multifarious elements in what Ingold (2011) calls the meshwork of flows is just as problematic as defining what 'nature' is. My use of the term 'nature' and its synonyms[1] apply to things or places other than those human-made. It is, however, apparent from my research that the complicated nature-human dichotomy is nebulous and my work chimes with much of Ingold's (1993, 2010) commentary on how relationships and actions or behaviour seem to forge the environment which *is* 'nature', and that as 'nature' is life, we (humans) *are* 'nature', we are part of every life that is 'nature' and not separate entities (Descola 2013).

For the participants of this study, being in the garden is a sensory experience that through numerous attentive eats allows life to be tasted, and during Covid, the garden provided some important reassurance that life keeps on living in times of turbulence. I explore four main foci: being with the garden, relationships with food, nourishment and the future of home gardening.

Being with the garden

The garden is home

This section explores how home gardening was perceived during the pandemic and attends to the ways that being in the garden are understood and enacted. Scant anthropological research exists on growing your own food (forthwith GYO) in UK *home-gardens*, as opposed to growing food on allotments[2], or in community or public green spaces. For other scholarship, the home-garden shares many of the same meanings associated with the concept of 'home' (Bhatti and Church 2001; Gross and Lane 2007). GYO in the home-garden enables people to engage with growing outside whilst being in an environment conveniently adjoined to the house so that any spare moment can be seized upon to pop outside and tend to a plant or a task. The most important distinction to make between GYO in allotments or other shared locations therefore, is that the home-garden is a private, personal, safe and unregulated home space (Bhatti and Church 2004).

The garden is a place to potter

All gardeners claim they enjoy 'pottering' in the garden, and whilst this somewhat British term is often used to refer to what gardeners do – they potter! – the definition of simply randomly 'doing things' is insufficient to capture the nuances in its use. Pottering conjures a sense of freedom, doing whatever you like in a private place, usually alone and lost in thought. For me, as for McGovern (2020), pottering is an opportunity to enjoy the meandering moment and settle myself among things by interacting with them, tending to them, feeling relaxed and safe. Importantly for the study's participants, pottering in the garden is a distraction from other routines or daily habits, and as Taliotis (2020, 22) explains, 'these trivial but absorbing tasks give your mind some much-needed time off from whatever issues and stresses it is straining to resolve – and that is incredibly important for wellbeing'.

Fig. 7.2 Pottering in the garden.

New to gardening, Rachel described the time she spent in the garden during Covid as pottering.

> It's very much just a sort of a weekend potter about type of a thing. It's not more than that. I don't think it ever would be.
>
> (Rachel)

Her comment suggests that she was unaware of the significance of this pottering behaviour. Having declared that she really 'wasn't herself' and needed to 'get out and do something', I understood her pottering to be a step she had taken to improve her mental health and a method that afforded an escape from other stressful areas of life.

The garden is an escape

> I'm just a typical gardener and I agree about mental health because I think that's why a lot of people have taken to the garden, because it's an area you can escape to and enjoy and just be lost in your own thoughts as well. It's definitely my happy space.
>
> (Jackie)

Jackie's comments link pottering with happiness and add the component of 'escaping' to being in the garden. Stuart-Smith (2020) says that people instinctively turn to nature at a time of crisis. During the Covid-19 lockdowns, the British people were told to stay at home (BBC 2020). People's lives and routines were severely disrupted, and restrictions were imposed on many daily activities such as shopping or meeting up with friends and going out away from home. Those working from home could not get away from the workplace (Mallett 2004) and people started to experience the home space as a confinement zone, and any feelings of comfort, safety and being able to potter contentedly no longer applied to the indoor home environment. Being in the house for the members of my research group felt like being caged or trapped. People talked of *escaping* to the garden, implying that even though they were still 'stuck' at home, they could nonetheless experience a release from their confinement. The garden as an escape, represented a place of freedom.

> It's like my little secret garden. It's lovely. Throughout the first lockdown I used to go down there once the kids had finished home schooling … I got to escape.
>
> (Amanda)

> [my garden] is an escape and it's very relaxing and … therapeutic … [Gardening] does tend to help you to forget that you know, there's a crisis out there … It really does sort of ease the emotional stress of lockdown.
>
> (Jackie)

I concur that when I am in the garden, I feel less stressed, less constrained, and more relaxed. Jung states that '[the answer] is not to escape, but to *do*' (Jung cited in Stuart-Smith 2020, 132 my emphasis). Indeed, the gardeners in this study went to the garden to *do* tasks, to grow their own food. Jackie and Rachel said they enjoyed the garden because it got them out and was something to do. The garden as a place fulfilled that need, and in alignment with Ingold's (1993) taskscape, participants understood gardening tasks as *doing* something to improve their own health and well-being as well as the health and well-being of the garden environment. These tasks deliberately or inadvertently forged relationships with the outdoor world and enabled gardeners to be more involved and immersed (Ingold 2010) in life all around them, including morethanhuman events such as the seasons and the weather.

The garden is a retreat

> [In] just a few minutes, and you kind of think, I'm here and I'm on my own … and yeah, it's just a happy place… there's something really personal about my veg patch.
>
> (Amanda)

> I can literally be transported as soon as I open the door and go outside.
>
> (Sharon)

What is it in the garden that engenders this feeling of freedom, or sense of calm that some cannot experience in the same way elsewhere?

Gardening may improve our physical health, because it involves being outdoors breathing fresh air, taking exercise and allowing our skin to absorb vitamin D. As such, gardening is increasingly recommended by physiotherapists for physical rehabilitation programmes after surgery or injury. Similarly, gardening is considered a treatment to improve mental health, with restoration therapies for people following trauma, suffering from PTSD, overcoming drug abuse, for refugees, or people severely and emotionally affected by events in their lives, often being held in gardens. Amanda and Eileen described their gardens as their sanctuary, and both discussed the importance of having *me-time* when gardening, agreeing it was necessary for their sanity. Sharon explained how gardening was good for her mental health, saying that physically going to the garden enabled her to be elsewhere in her mind too.

The garden nurtures relationships

The need to establish and nurture relationships is an essential part of life. During the Covid-19 lockdowns, relationships altered as participants' interactions, other than with those they lived with, tended to rely on digital media – phone, computer, internet. Accounts from gardeners suggest that people wanted interaction directly with living things, not via a screen, and gardeners seemed to achieve this by nurturing their relations in the garden as an alternative to human contact, just as with horticultural therapy. Accordingly, people gardened to satisfy a need for their senses to engage with something or some place that was lively or agential (Wilson 1990) and these phenomenological experiences were shared with otherthanhuman entities thereby nourishing a sense of wider existential interconnectedness.

Gardening: Attention and sensorium

As Ingold (2011) affirms, the world is not just something to look at, but something to experience, and immersion in the world is the way to experience it (see also Dickinson 2013; Whitehouse 2017). Such an immersion might then necessitate a move away from screens and an increased regard for how sentient we are as humans (Cohen n.d.). Crowther (2013) and Counihan (2015) explore the increasing distance between humans and their ability to give unhurried sensory attention to their lives, and question how urban contemporary lifestyles impact on how living is experienced. In association, access to green spaces is in decline, particularly among the urban populations of our industrialised and consumer-oriented nations, despite public media identifying that there is a need for this. Whilst technological advancements extend and proliferate our connections across vast distances, these changes increasingly occupy people's days and make it difficult to find the time for other forms of connection. Helen alludes to this distancing in her abhorrence of current trends to offer 'substitutes' for *living* landscapes, 'Why are we concreting everything?... I mean, Astroturf, it's bloody plastic!' Connections are not formed by just *being* outside nor in a facsimile of nature.

The Summer 2021 edition of the National Trust's magazine for members (Bevan 2021, 21) featured a cover article entitled 'Escape into a Garden', and referred to conservationists' use of the expression 'spirit of place' to describe the qualities which make a place special. This specialness was described as an experience, that has 'the power to absorb you,' (invoking the term pottering once again), '[as gardens] touch all the senses' – enabling an immersion that can significantly affect our lives (Bevan 2021, 21).

> I feel very much at one with nature ... it [GYO] sort of ticks a box when it comes to you know, the basic instinct side of things, [*pause*] nurturing myself, you know.
> (Jackie)

It was not easy sometimes for participants to articulate precisely how or why they feel better in the garden and their reasons for GYO. A lexicon for explaining our sensory responses is lacking in English, alluded to by Pollan (2013) and described by Goodey (1973, 4) as problematic:[3]

> when perceiving his [sic] environment, an individual does not sub-divide his observations by disciplinary categories, he views the totality 'out there'... Perception is therefore an extremely dynamic process and one which is very difficult to monitor.

Participants in my study wanted to describe something that I believe is not just an interface of agential happening, but a holistic response to a plethora of stimuli, just as Goodey describes. Having recorded myself potting-on some seedlings in the greenhouse and having attempted to provide a commentary for the film by talking about my phenomenological encounters, I can confirm how difficult it is to identify, talk about and focus on any one of my senses. I was absorbed in my task and felt an overwhelming sense of immersion in the world without any cognitive awareness, yet I was aware of everything nonetheless – the warmth of the sun, the sounds of the birds, the rustle of the wind, the scent of the soil, the fragility of the seedlings and so on. All these stimuli were felt at the same time, and it was impossible to prise them apart and talk about any one of them without being disingenuous to the other in acknowledging their contribution. I developed a greater depth of understanding of synaesthesia having felt it myself and more so, when recognising that's what *it* is.

Shilling (2017) notes difficulties in explaining the difference between how someone thinks and how someone feels and explores how the senses inform the body separate to cognitive acknowledgement. Stuart-Smith (2020) posits that a difficulty in expressing sensory data could be the reason for relatively little research into the beneficial effects of gardening to date. Cohen's (n.d.) discourse on not five, but fifty-four senses talks of the languageless of nature, an anthropocentric statement implying that without language, other entities cannot communicate. The fact that we as humans do not share the same language as others, does not mean that living things cannot communicate within and among themselves (Sheldrake 2020) and there is growing interest in this field among scholars. In fact, an expanding area of research in ecological and nutritional anthropology is on plants' agency on humans surmising that hunter-gatherers facilitate their (plants) (re)growth the following year (Crowther 2013; Attala 2017).

The challenge then, is for humanity to respond to the sensory organs that we evolved with and pay attention to how the world communicates. Being in the garden enables us to do this. As Helen said in my fieldwork:

> [Gardening] is almost like a meditational thing, like a re-connection sort of, with your, your you know, inner goings-on.

Relationships with food

> Food is our everyday creative and meaningful engagement with nature through culture.
>
> Crowther (2013, xviii)

Our inner 'goings-on', as Helen calls it, brings me to the subject of gardening as a means of sourcing food: the ultimate connection between us and the outside world is when we ingest food, enacting a symbiosis (Attala 2017; Gilbert 2017). We bring the world into our bodies, and it becomes us – an assimilation (Beeton 1890; Kass in Bennett 2010).

The abrogation

Trust, provenance and quality of foodstuffs are being questioned as the globalisation of cultures, including food tastes and choices, becomes more homogeneous (Clark 2004). Scholars increasingly document the separation between people and their food origins (Mason 2004; Counihan 2008; Crowther 2013). Coveney (2014, 2) notes that our current foodways are '… abrogating our relationship with food' as we increasingly rely on others to produce and provide food, and Clark (2004, 418) declares people are 'increasingly alienated from that which keeps them alive'.

Unlike any other European country, five supermarkets dominate the UK's food supply (Crowther 2013), dictating the cosmetic appearance of produce, insisting on uniform sizes, shapes, and weights to aid automated packaging and labelling systems, all to maximise efficiency and profit and enable consumers to have out-of-season foods all year round. To meet these demands, farmers have restricted the variety of crops grown to those that ensure yield and resist disease, flooding the market with limited choices, bland produce (Luetchford, 2014) and, according to de Certeau and Giard (1998), mediocrity. Plants grow in seasons and offer variety but the modern food supply system attempts to negate the seasons and exert power and dominance, not only on the growers, but on the natural cycle of life (Brooks et al. 2016). The increasing distance between us and our food is also evidenced with the statistic that over 80 per cent of Britain's food is imported (Edwards 2019).

The current normative food systems preferred in Euro-American cities 'inform(s) the debate on how we might question and transform a society defined by constant busyness' (Schoneboom 2018, 374) and one driven by economics, not the health and well-being of the planet and its inhabitants (Verhaeghe 2014). I share these views and add that people do not necessarily need food that is convenient i.e., easy to prepare, cook or eat, because they have less time to do these things: people are choosing instead to spend their time doing *other* things. Gardeners, however, invest their time and money in GYO because they believe it is an important lifestyle choice. Gardeners' decisions to grow at least some fresh produce are indeed an attempt to oppose the commercialisation of the food industry.

> It's all about balance, though, isn't it? … We're always trying to push something forward or take something back or whatever. We're always trying to manipulate things. It's the same with you know, like growing vegetables out of season by hydroponics and all things like that … for manufacturers profit probably. You know, we are forgetting what it's like to sort of look forward to something.
> (Sharon)
>
> I think seasons are there for a reason …
> (Liz)

The acts and tastes of growing

The act of growing food, often represented as a unilinear process where choices are initiated by the gardener, is in fact a complicated set of relationships enacted between multiple players. Atchison et al. (2010) explore the complex interactions between gardeners and plants, asserting that plants are just as much agents as the gardener. This invokes Attala's (2017, 2019) Edibility Approach, which homes in on the relationship that plants instigate with humans (rather than, or as well as the other way around) during, before and after digestion. Attala's work suggests that plants taste wonderful for a reason.

Part of their mode of survival is to be attractive to animals so that they revisit and encourage seed dispersion or cultivation. However, monoculture and industrialised processing of foodstuffs indisputably destroys natural flavour and with the number of vegetable varieties grown reducing, plants' way of communicating messages (by tasting good), is ignored.

Gardeners place great importance on growing a range of varieties:

> I've taken the decision [to GYO] mainly for the environment ... and for me it's about the quality and not quantity ... I absolutely just love vegetarian ... that's another reason why I'm starting to grow more of my own because there are certain particular vegetables and stuff that you can't get in the supermarket and if you're eating predominantly veg, you want to have a bit of variety don't you? More tastes and flavours ...
> (Helen)

Participants found it difficult to understand how people could be satisfied with the taste of produce bought from the supermarkets. Liz and Jackie both remarked that unless people have tasted freshly harvested produce, they would not know the difference. According to the gardeners I spoke with, environmental concerns linked with taste were the main drivers that motivated them to grow their own food, which suggests that Attala's (2017) claim that edible plants entice and seduce people through their flavour is upheld.

Bourdieu (1979) notes that taste is the most indelible of our senses stimulated at a very early age. Gardeners maintain and regularly exercise their taste buds, dismissing bold statements that the palate is a dead organ and that people have a reduced ability to respond to their senses (Orwell 1959). Many scholars support the notion that taste triggers deep memories and can transport you to another time (Counihan 1984; Sutton 2001). Taste triggers far more than a sensory gland's response and is impacted by the experiences associated with that food (Goodey 1973; Sutton 2001). Crowther (2013, 131) attests, 'The smell and taste of a dish are capable of transporting us back in time and place, reconnecting us to our memories of people and shared meals'. Dahl (2010) describes how borscht embodies place for her, 'To me, it's Norway in a bowl'.

PLANTS MATTER   173

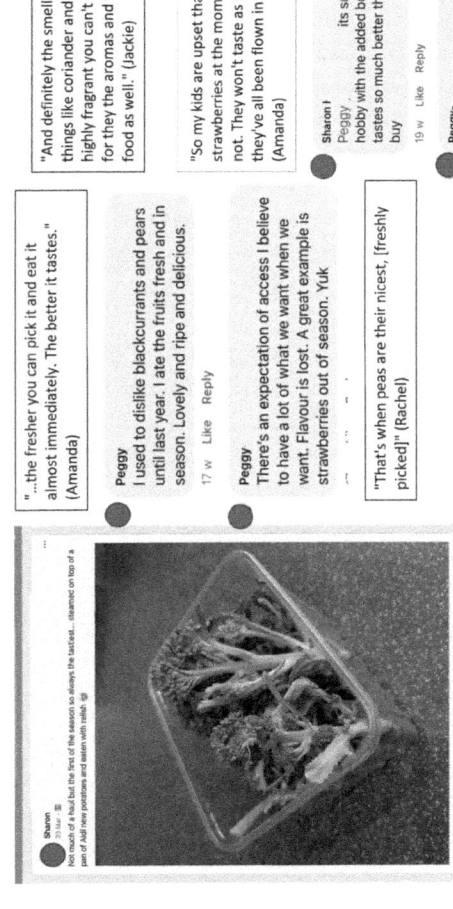

Fig. 7.3 The unsurpassed taste of home-grown produce.

If memories of the past can affect taste in such a way, then the comments collected during fieldwork suggest that the memories of food's *growth* are also embodied in food and could affect taste in the same way. A superior taste, therefore, is experienced with memories of its (the food's) creation.

Another observation regarding taste suggests that anticipation plays a part. Research shows that the longer you wait for food, the tastier it is when you eventually eat it (Jenkinson 2020). Whilst the research may only explore a short waiting period, it is telling that waiting for the first fresh pickings of the season conjures a much better taste, enjoyment, and satiation than any readily available produce from a shop. The benefits of the cephalic stage of digestion are extolled by nutritionists and alternative health practitioners (Lane 2017; Williams 2017) yet, I have struggled to find studies on the longevity of this stage. I maintain that GYO contributes to this important aspect of eating because of the many meanings ascribed to the food's coming-into-being, and the relationships forged with it during its nurture and growth.

In addition to experiencing the delightful flavour of seasonal produce, gardeners attempt to sustain the quality of taste throughout the year by preserving their food. Jackie regularly batch-cooks and freezes meals enabling her to enjoy the fresh taste of home-cooking at any time of the year and on busy days. Liz and Helen also freeze surplus harvests:

> [I] froze some runner beans last year. [We] still got that taste … They were still frozen but nothing like them horrible soap-tasting things and chopped up veg rubbish people buy but they actually were really tasty and tasted of beans still.
> (Liz)

> That kind of like making things last longer, storing things so you got a bit of a taste of, 'specially in the deep depths of winter. You've got like a nice taste of spring and summer to sort of like hark back to. There's something quite sort of primaeval about that, in a way. Do you know what I mean? Like squirrelling stuff away … I enjoy that aspect of, you know, gardening and foraging as well.
> (Helen)

Foraging flagged up the theme of taste also. Helen told us about her day foraging for wild garlic:

> I find it actually a very sort of mindful and relaxing thing to do. I think being out in nature, whether it's in your garden or a forest or wherever, it may be just that you know, being there for the purpose of getting food is quite … it fulfils you on a lot of levels I think and I really enjoy it very much … it gives you that sort of feeling of wellbeing.

Foraging is not of course, GYO. However, all the gardeners I spoke with had stories to tell about foraged foods, highlighting how they respond to the seasons, and how they sought not only the smells and tastes of freshly picked food, but the experience of harvesting it as well. This notion also reminds me of Liz's comments lamenting the lack of Pick-Your-Own farms these days, as she recalled the annual excitement of a day out to pick as many strawberries and raspberries as she could and the various produce subsequently made with the hoard.

People want a relationship with their food akin to any other relationship they may have with people or animals. As Jackie suggests, 'I think our basic instinct is to nurture, and maybe gardeners do this more than any others'. The gardener's relationship with food might manifest in care for its 'homelife' (origin); consideration for how it grows; respect for its role in the complex ecosystem of life and within our own bodies (variety, diet); dedicating time to its preparation (harvest, peeling, chopping, cooking, consumption). There are many studies on relationships with food, however there are few on the relationships that emerge from the experiences *when growing* the food in a home-garden, nor how this couples with the experience of preparation, cooking, consumption and then disposal of food and its regrowth (from composting to saving and sowing seed). My study's gardeners rejoiced in being part of food's life, from seed to plate to compost and not only enjoy the pleasure of eating, but are satiated by an eating *experience* that cannot be obtained from food that has been sourced by someone else.

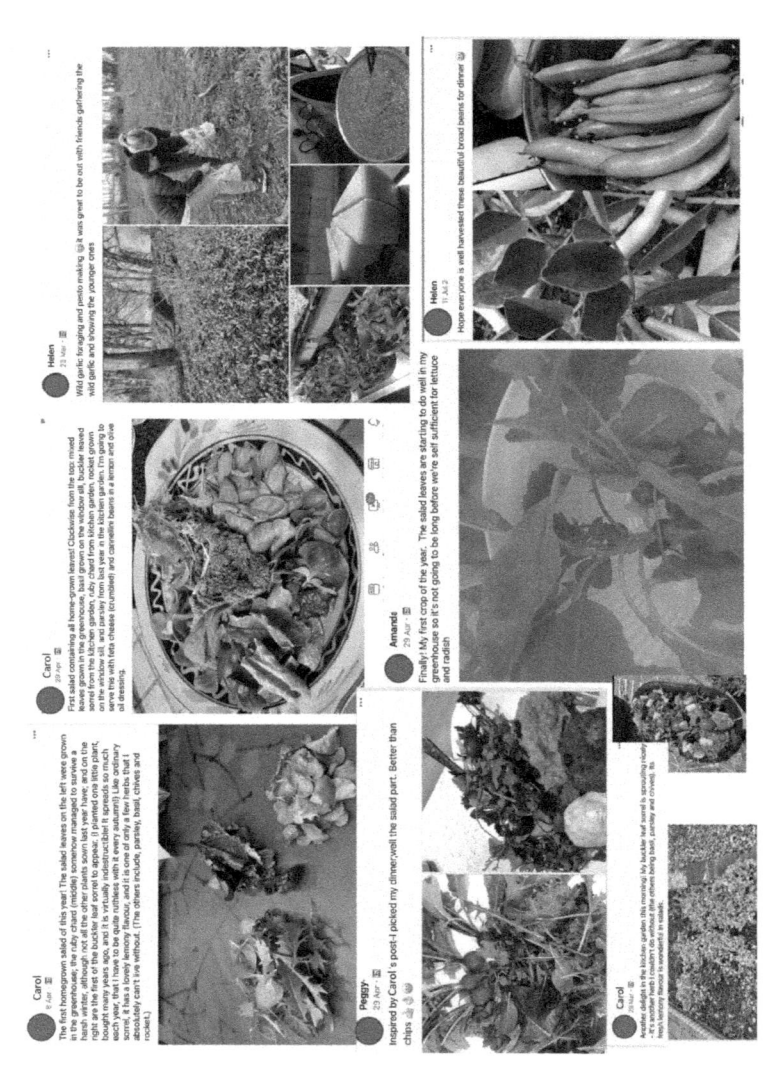

Fig. 7.4 Harvesting salads and foraging for wild garlic.

Nourishment

Nutritionists have yet to jump on the band wagon and extol the virtues of GYO to reach the ears of global policy makers. They hold to nutritional statistics that eating 'fresh' vegetables and fruit is the way to a healthier life even if these are acquired through our normative foodways[4].

Nutrition

Nutrient values printed onto food labels may allude to the quality of produce. However, whilst the average consumer may develop a nutritional consciousness as a result of government guidelines and marketing, nutrition as a science has a number of flaws. Firstly, as Coveney (2000) points out, nutrition science objectifies bodies and applies recommended values for specific outcomes on hugely diverse populations whilst ignoring the uniqueness of individuals' size, shape, internal systems, quantities, qualities, thoughts, and feelings. Secondly, data in the UK (Department of Health 2013) on nutrient values of fruit and vegetables comprises a list of average figures, calculated from composite samples of foods harvested at different times of the year and from different locations, with different packaging, storage, transportation, and treatment histories. Consequently, the figures do not represent the *actual* values in a particular product. Vitamins are labile (Coveney 2000).[5] Taste may be a signifier here, because as the gardeners in my study affirm, tastes are stronger just after picking. It would seem therefore, that if the taste is lacking, the nutrients are too. From this, I deduce that taste is an indicator of quality, yet, bizarrely Atwater (Shapiro, cited in Coveney 2000), who devised nutritional science, said that taste plays no part in healthy food! Thirdly, and of most significance for anthropologists, because eating is much more than just obtaining nutrients, the phenomenological and visceral experiences of human life, culture and meanings, are missing with nutritionism. Consequently, policies and food regulations informed by nutritional values of individual components ignores the human need for a relationship with food.

178 PLANTS MATTER

Meaning-making

Debevec and Tivadar (2006) posit that homemade is full of goodness because social, cultural, and visceral experiences create a relationship with food through cooking, preparing, and eating it. My study implies that home*grown* is another factor to contribute to the goodness of home*made*, bringing with it all embodiments from gardening.

I created a *Home-Grown Food Vortex* illustration below, to show how the phenomenological and visceral experiences of GYO contribute to meaning and to taste, based on comments shared during my research. I drew this because participants struggled to articulate their connections with the plants, and I was sure that representing this in a diagram might address this, since vocabulary was not required, although paradoxically, I used words in the illustration! It became apparent, however, that interlocutors felt their connections with the

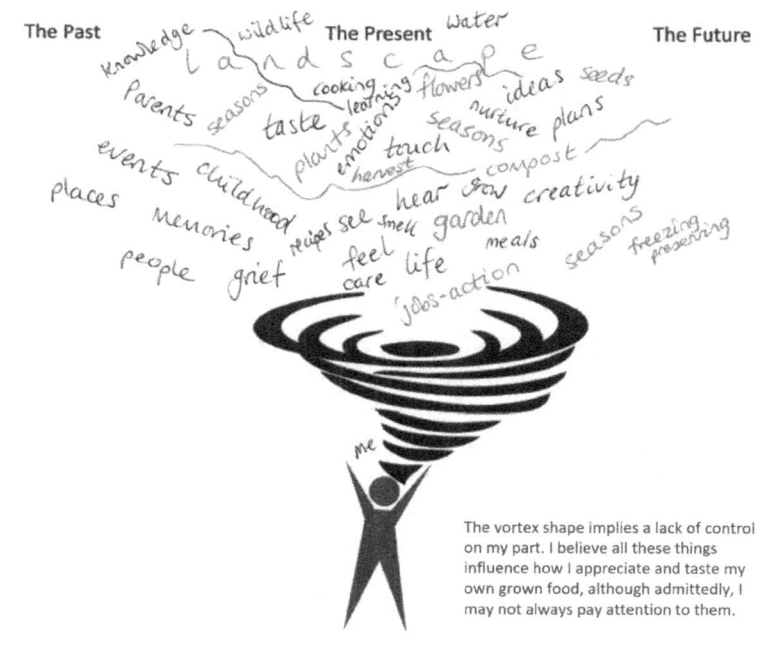

The vortex shape implies a lack of control on my part. I believe all these things influence how I appreciate and taste my own grown food, although admittedly, I may not always pay attention to them.

Fig. 7.5 My Initial Vision of the Embodiments of Gardening – The Home-Grown Food Vortex, as the gardener and eater.
Author's own, 2021.

food they grew formed a multi-dimensional flow and on reflection then, the vortex graphic does not accurately represent gardening embodiments in food because it portrays me, the human gardener, as a passive receiver of these various 'inputs'.

My attempt to reshape the vortex model confirmed how complex the holistic food-getting process is and I concede that the connections with food that gardeners grow, together with the experiences and relations they nurture in the garden during its growth, are far more nebulous than anything I could portray in a drawing.

Ingold's (2010) immersion in the flux of life emerges once again here, also highlighted by Sharon's musings on an attentiveness to her senses:

> it's all one big sort of like web and we are all connected and I think that's why I tend to try and avoid out of season stuff ... cos you know, my body is working properly, and I know what I need through the year, and that's what I'm trying to eat.

Sustainability

All the gardeners I encountered hold strong ethics of sustainability and stewardship for the planet and see themselves as caretakers not just of the plants they grow but of the environment in which they might flourish. No more is the speed of change witnessed in the landscape than in our current planetary climate crisis. Current foodways significantly contribute to it (Mason, 2004) and in line with Descola (2013) ousting a nature/human dichotomy, we cannot separate human health and the health of the planet because both impact on each other. Participants expressed outrage at extensive food miles and their impact on the environment.

> Why are we flying stuff in from Brazil and Morocco? Why not just eat seasonal veg the way you are supposed to eat it? ... It kinda detracts from the seasons you know? Like British asparagus at its best straight outa the field, straight onto the plate. It should be celebrated and enjoyed when in summer here ... you should look forward to it, whereas now you can just get anything anytime.
>
> (Liz)

In maintaining a proximity to their foods' origins, gardeners are *acting* to do their bit, however small. Jackie's explanation as to why she grows her own vegetables, typically represents all my participants' views:

> it just gives you such a feel-good buzz when you've picked everything that you've grown that you've put your labour and heart into growing. And it looks lovely in the basket. It smells good and the food that you produce you know tastes good and is so much better than what you would buy in a supermarket. There's no processing factors to take into consideration. You're eating something that's good and healthy and nutritious ... that feeling ... [*pause*] that ... [*pause*] you've reduced your footprint, your carbon footprint a little bit, and you know you've not used the Earth's resources in transporting things or packaging them or whatever, so that, you know ... [*pause*] that feeling of doing your little bit for the planet as well. It's all important for me.

The future of GYO

England is a 'nation of gardeners' (Willes 2014), yet over 3 million more people took up gardening during the pandemic, including those with limited budgets, taking the total number of gardeners to nearly 47 per cent of the UK's population (HTA 2021). The impact of this number of gardeners growing their own food, not only on food security, but also on human health and well-being and on the health of the planet, could become significant. The pandemic coincided with increasing questions about food supplies (Latham 2021). With empty shelves, panic-buying and everyone's daily routines upset with the stay-at-home order, Helen pondered:

> You know, with the food shortages etc. in our sort of you know developed, nice little bubble part of the world, people have never really had to even think about where they shop, you know, where their food comes from.

The Covid-19 lockdowns in England were not the only provocation for people asking more questions about the origins of their food. The UK's withdrawal from the European Union, which coincided with the pandemic during 2020, also impacted food supplies and continues to do so. However, if GYO were to be encouraged and recognised as a significant option for food-provision, several issues might arise.

The first, flagged by Sara Venn of Incredible Edible Bristol argues, 'If we are telling people to garden because it's really good for you, … we need to create those spaces for people'. Gardeners in my study are not able to imagine being without a garden now that they've experienced it, yet, the number of UK households with access to a garden is declining (HTA 2021) and there are no policies that require provision of a garden or any outdoor space in new house building guidelines in the UK (NHBC 2021). Also, as part of the UK Government's 25 Year Environment Plan, in 2019, a Year of Green Action was promoted, which suggested ways to use your garden to make 'greener choices', yet GYO was not listed as one of the ideas.

Secondly, allaying concerns about the scale of GYO practices, I refer to examples from Tanzania, where gardens are designed for 'ongoingness' rather than scalability (Langwick 2018), a view that gardeners share all over the world anyway (Barthel, Parker and Ernstson 2015; Koczberski et al. 2018). Gardeners, it seems, do not want to scale up and provide for *everyone* but are more than content to provide for themselves and their own family/friends' needs, nurturing relations between them and their food, including their relationship with the green outdoors and all the ethical associations of sustaining their food supply (Vávra, Smutná and Hruška 2021).

Thirdly, priorities are a vast, subjective concept for further research, although there are many examples of small-scale societies who value food-getting work highly because it is essential for life (Crowther 2013). For Andalusian field labourers, the people who produce food are first in the hierarchy of things for the reason cited above (Pratt 2014). In line with Eriksen's (1995) comments on the division of labour among Mundurucú in the Amazon, gardening as food-provisioning is more important than hunting. During the pandemic, those in the food industry were identified as key workers and now, so many more people are asking about food and what is essential for *their* lives.

I posit that GYO is essential for life because individuals become responsible for their own food sourcing (Bray 2016) and nurture a relationship with it that enhances its quality (Luetchford 2014). Seymour

(2002, 13) notably declares that for retailers in today's global food system, quality is determined by its long shelf life and he calls such food 'dead: all the life has been taken out of it'. He adds that GYO means 'striving for a *higher* standard of living, not going back to the old days, or a *lesser* lifestyle' (my emphasis).

In conclusion, Spence's work (Fleming 2014) on how our senses give us clues that affect our tastes, has parallels in how the embodiment of the food taskscape affects how satiated or nourished we are by our food. My research suggests that the taskscape of the garden for food-sourcing, the meanings ascribed to the actions within it, the phenomenological and visceral experiences that occur in the garden and the fusion of all time dimensions around it (memories, re-enactments, non-human seasonalities and events), constitute an enhanced quality of nourishment from the food that is grown there. Ingold's (2007, S30, 2011) theory conceives that the weather is not an object, but an *experience*. He dallies with interchanging nouns and verbs, elucidating that the 'wind *is* the blowing' just as the 'fire *is* its burning'. Applying this theory to GYO, nourishment *is* the home-growing. Eating home-grown food according to gardeners, with all its associations and embodied meanings, nourishes more than any substance that is produced, supplied, bought, manufactured, or grown by someone else, is uniquely experienced by each and every individual, and positions the human inseparable and immersed in the flux of life on earth.

This chapter centres on my own research conducted during the Covid-19 lockdowns in 2020, when I carried out multi-sited fieldwork with gardeners growing their own food in their home-gardens, and draws on much of Ingold's (1993, 2002, 2007, 2010, 2011) work on immersion and the entanglements of life. It is hoped by many that the pandemic will be a springboard for change (Keith-Lucas 2020) and my study not only contributes to the understanding of what it means to grow your own food but also adds to the global discourse on health and well-being, food sovereignty and even the climate crisis. I have explained how the concept of home, as a safe and uncensored place, extends to the home-garden, where gardeners grow their own food (GYO) and how the garden itself holds multifarious meanings and associations that affect well-being. I have explored how gardeners' innate connection, or rather immersion in the natural world, accrues meaning for the food consumed, and how those who garden act on their sustainability ethics by growing their own, deploring the UK's current foodways. I detail the significance of taste in establishing quality, and link this to the agency of the 'natural' world, into and within which we, as humans, are not separate to, but part of. I have suggested a new appreciation of nourishment: that a greater satiation is achieved by eating food that individuals have grown or foraged themselves. Therefore, in emulating one of Ingold's (2007) phrases, I believe that nourishment *is* the home-growing – an immersion in the life of the food I consume, the life of which becomes and is, me.

References

Atchison, J., Head, L. and Gates, A., 2010. 'Wheat as food, wheat as industrial substance: Comparative geographies of transformation and mobility', *Geoforum* 41(2), 236–46, doi: *10.1016/j.geoforum.2009.09.006*.

Attala, L. 2017. '"The Edibility Approach": Using edibility to explore relationships, plant agency and the porosity of species' boundaries', *Advances in Anthropology* 7(3), 125–45, doi: *10.4236/aa.2017.73009*.

Attala, L. 2019.'"I am Apple": relationships of the flesh. Exploring the corporeal entanglements of eating plants in the Amazon', in L. Attala and L. Steel (eds), *Body Matters: Exploring the Materiality of the Human Body*. Cardiff: University of Wales Press, pp. 39–61.

Barthel, S., Parker, J. and Ernstson, H., 2015. 'Food and green space in cities: A resilience lens on gardens and urban environmental movements', *Urban Studies* 52(7), 1321–38, doi: *10.1177/0042098012472744*.

BBC, 2020. *Coronavirus*. [online] Available at: *https://www.bbc.co.uk/news/coronavirus*. Accessed September 2021.

Beeton, I., 1890. *Mrs Beeton's Household Management*. Second edn. London: Ward, Lock and Co. Ltd.

Bennett, J., 2010. *Vibrant Matter: A Political Ecology of Things*. Durham, NC and London: Duke University Press.

Bevan, R., 2021. 'Escape into a garden', *National Trust Magazine*, 20–7.

Bhatti, M. and Church, A., 2001. 'Cultivating Natures', *Sociology*, 35(2), 365–383.

Bhatti, M. and Church, A., 2004. 'Home, the culture of nature and meanings of gardens in late modernity', *Housing Studies* 19(1), 37–51.

Bourdieu, P., 1979. *Distinction A Social Critique of the Judgement of Taste*, trans. R. Nice. 1984. Abingdon: Routledge Kegan and Paul.

Bray, F., 2016. 'Fooding Farmers and feeding the nation in modern Malaysia: The political economy of food and taste', in J. A. Klein and J. L. Watson (eds), *The Handbook of Food and Anthropology*. London: Bloomsbury Academic, pp. 173–99.

Brooks, S., Burges Watson, D., Draper, A., Goodman, M. K., Kvalvaag, H. and Wills, W., 2016. 'Chewing on choice', in E. J. Abbots and A. Lavis (eds), *Why We Eat, How We Eat. Contemporary Encounters between Foods and Bodies*. Abingdon: Routledge, pp. 149–68.

de Certeau, M. and Giard, L., 1998. 'The Nourishing arts', in C. Counihan and P. van Esterik (eds), *Food and Culture: A Reader*. Second edn. New York, NY: Routledge, pp. 67–77.

Clark, D., 2004. 'The raw and the rotten: punk cuisine', in C. Counihan and P. van Esterik (eds), *Food and Culture: A Reader*. Second edn. New York, NY: Routledge, pp. 411–22.

Cohen, M. J., n.d. *Project NatureConnect*. [online] Available at: *http://ecopsych.com/insight53senses.html*. Accessed January 2022.

Counihan, C., 2008. 'Mexicanas' food voice and differential consciousness in the San Luis Valley of Colorado', in C. Counihan and P. van Esterik (eds), *Food and Culture: A Reader*. Second edn. New York, NY: Routledge, pp. 354–68.

Counihan, C., 2015. 'Ethnography of farmers' markets: Studying culture, place, and food democracy', in C. Lowe Swift and R. Wilk (eds), *Teaching Food and Culture*. Walnut Creek, CA: Left Coast Press, pp. 113–28.

Counihan, C., 1984. 'Bread as world: Food habits and social relations in modernizing Sardinia', *Anthropological Quarterly* 57(2), 47–59.

Coveney, J., 2000. *Food, Morals and Meaning: The pleasure and anxiety of eating*. London: Routledge.

Coveney, J., 2014. *Food*. Abingdon: Routledge.

Crowther, G., 2013. *Eating Culture: An Anthropological Guide to Food*. Plymouth: NBN International.

Dahl, S., 2010. *Beet Soup*. [online] Available at: *https://www.cookstr.com/recipes/beet-soup-sophie-dahl*. Accessed 20 January 2019.

Debevec, L. and Tivadar, B., 2006. 'Making connections through foodways: contemporary issues in anthropological and sociological studies of food', *Anthropological Notebooks* 12(1), 5–16.

Department of Health, 2013. *Nutrient analysis of fruit and vegetables, sampling report*. [pdf] London: Department of Health. [online]

Available at: https://www.gov.uk/government/publications/nutrient-analysis-of-fruit-and-vegetables. Accessed January 2022.

Descola, P., 2013. *Beyond Nature and Culture*. Trans. J. Lloyd. London: University of Chicago Press, doi: 10.1521/prev.2017.104.4.485.

Dickinson, E., 2013. 'The misdiagnosis: Rethinking "nature-deficit disorder"', *Environmental Communication*, 7(3), 315-35, doi: 10.1080/17524032.2013.802704.

Edwards, J., 2019. *Say goodbye to tea and carrots: 80% of British food is imported so there will be food shortages if there's a no-deal Brexit, HSBC tells clients*. [online] Available at: https://www.businessinsider.com/no-deal-brexit-percentage-british-food-imported-shortages-2019-1?r=US&IR=T. Accessed January 2022.

Eriksen, T. H., 1995. *Small Places, Large Issues: An Introduction to Social and Cultural Anthropology*. London: Pluto Press.

Fleming, A., 2014. *Charles Spence: Food Scientist Changing The Way We Eat, The Guardian online*. [online] Available at: https://www.healthline.com/health/grounding. Accessed January 2021.

Gilbert, S. F., 2017. 'Holobiont by birth: Multilineage individuals as the concretion of cooperative processes', in A. L. Tsing, H. A. Swanson, E. Gan and N. Bubandt (eds), *Arts of Living on a Damaged Planet*. Minneapolis, MA and London: University of Minnesota Press, pp. M73–M90.

Goodey, B., 1973. *Perception of the Environment: An Introduction to the Literature*. Birmingham: University of Birmingham.

Gross, H. and Lane, N., 2007. 'Landscapes of the lifespan: Exploring accounts of own gardens and gardening', *Journal of Environmental Psychology* 27(3), 225–41, doi: 10.1016/j.jenvp.2007.04.003.

Harper, K. and Afonso, A. I., 2016. 'Cultivating civic ecology: A photovoice study with urban gardeners in Lisbon, Portugal', *Anthropology in Action* 23(1), 6–13, doi: 10.3167/aia.2016.230102.

HTA, 2021. *Millions of new British gardeners take root as a result of lockdown*. [online] Available at: https://hta.org.uk/news-current-issues/news-current/news/millions-of-new-british-gardeners.html. Accessed January 2021.

Ingold, T., 1993. 'The temporality of the landscape', *World Archaeology* 25(2), 152–74.

Ingold, T., 2002. *The Perception of the Environment: Essays on Livelihood, Dwelling and Skill*. London and New York, NY: Routledge.

Ingold, T., 2007. 'Earth, sky, wind, and weather', *The Journal of the Royal Anthropological Institute* 13, S19–S38, doi: 10.1017/S0001972000085715.

Ingold, T., 2010. 'Bringing things to life: Creative entanglements in a world of materials', *Realities working papers* 15, pp. 1–14, https://eprints.ncrm.ac.uk/id/eprint/1306/1/0510_creative_entanglements.pdf.

Ingold, T., 2011. *Being Alive. Essays on Movement, Knowledge and Description*. Abingdon: Routledge.

Jenkinson, A., 2020. *Why We Eat (Too Much) – The New Science of Appetite*. Ebook. London: Penguin Books Ltd.

Keith-Lucas, S., 2020. *Climate Check*. [online] Available at: https://www.bbc.co.uk/weather/features/55280683. Accessed: 4 February 2021.

Koczberski, G., Curry, G. N., Bue, V., Germis, E., Nake, S. and Tilden, G. M., 2018. 'Diffusing Risk and building resilience through innovation: reciprocal exchange relationships, livelihood vulnerability and food security amongst smallholder farmers in Papua New Guinea', *Human Ecology* 46(6), 801–14, doi: 10.1007/s10745-018-0032-9.

Lane, E., 2017. *Understanding Digestion: The Cephalic Phase, Integrative Health Education*. [online] Available at: https://www.integrativehealth.co.uk/understanding-digestion-the-cephalic-phase/. Accessed January 2022.

Langwick, S. A., 2018. 'A politics of habitability: Plants, healing, and sovereignty in a toxic world', *Cultural Anthropology* 33(3), 415–43, doi: 10.14506/ca33.3.06.

Latham, K., 2021. *Has coronavirus made us more ethical consumers?* [online] Available at: https://www.bbc.co.uk/news/business-55630144. Accessed 30 April 2021.

Luetchford, P., 2014. 'Food and consumption', in J. Pratt and P. Luetchford (eds), *Food for Change*. London: Pluto Press, pp. 47–70.

Mallett, S., 2004. 'Understanding home: A critical review of the literature', *Sociological Review* 52(1), 62–89, doi: 10.1111/j.1467-954x.2004.00442.x.

Mason, L., 2004. *Food Culture in Great Britain*. Westport, CT: Greenwood Press.

McGovern, A., 2020. *Pottering: A Cure for Modern Life*. London: Laurence King Publishing.

NHBC, 2021. *NHBC Standards 2021*. [online] Available at: https://www.nhbc.co.uk/builders/products-and-services/techzone/nhbc-standards/standards-2021. Accessed January 2022.

Orwell, G., 1959. *The Road to Wigan Pier*. 2001st edn., ed. P. Davison. London: Penguin Classics.

Pollan, M., 2013. *Cooked: A Natural History of Transformation*. London: Allen Lane.

Pratt, J., 2014. 'Farming and its values', in J. Pratt and P. Luetchford (eds), *Food for Change*. London: Pluto Press, pp. 24–46.

Schoneboom, A., 2018. 'It makes you make the time: "Obligatory" leisure, work intensification and allotment gardening', *Ethnography* 19(3), 360–78, doi: 10.1177/1466138117728738.

Seymour, J., 2002. *The New Complete Book of Self-Sufficiency*. London: Dorling Kindersley Limited.

Sheldrake, M., 2020. *Entangled Life: How Fungi Make Our Worlds, Change Our Minds and Shape Our Futures*. London: Bodley Head.

Shilling, C., 2017. 'Body pedagogics: Embodiment, cognition and cultural transmission', *Sociology* 51(6), 1205–21, doi: 10.1177/0038038516641868.

Soga, M., Gaston, K. J. and Yamaura, Y., 2017. 'Gardening is beneficial for health: A meta-analysis', *Preventive Medicine Reports* 5, 92–9, doi: 10.1016/j.pmedr.2016.11.007.

Strang, V., 2006. 'Substantial connections: Water and identity in an English cultural landscape', *Worldviews: Environment, Culture, Religion* 10(2), 155–77, doi: 10.1163/156853506777965820.

Stuart-Smith, S., 2020. *The Well Gardened Mind: Rediscovering Nature in the Modern World*. London: William Collins.

Sutton, D. E., 2001. *Remembrance of Repasts*. Oxford: Berg.

Taliotis, X., 2020. 'Pottering delights', *Breathe*, 22–3.

Thompson, R., 2018. 'Gardening for health: A regular dose of gardening', *Clinical Medicine, Journal of the Royal College of Physicians of London* 18(3), 201–05, doi: 10.7861/clinmedicine.18-3-201.

Vávra, J., Smutná, Z. and Hruška, V., 2021. 'Why I would want to live in the village if I was not interested in cultivating the plot? A study of home gardening in rural Czechia', *Sustainability* 13(2), doi: 10.3390/su13020706.

Verhaeghe, P., 2014. *What About Me? The Struggle for Identity in a Market-Based Society*. Trans. J. Hedley-Prole. Victoria, Australia: Scribe Publications Pty Ltd.

Wesselow, M. and Mashele, N. J., 2019. '"Who needs money if you got hands, if you got plants" forming community resilience in two urban gardening networks in South Africa', *Human Ecology* 47(6), 855–64, doi: 10.1007/s10745-019-00116-5.

Whitehouse, A., 2017. 'Loudly sing cuckoo: More-than-human seasonalities in Britain', *Sociological Review* 65(1_suppl), 171–87, doi: 10.1177/0081176917693533.

Willes, M., 2014. *The Gardens of the British Working Class*. London: Yale University Press.

Williams, N., 2017. *It's Not You, It's Your Hormones. The Essential Guide for Women over 40 to Fight Fat, Fatigue and Hormone Havoc*. Bramley: Practical Inspiration Publishing.

Wilson, E. O., 1990. *Biophilia: The Human Bond with Other Species*. Cambridge, MA: Harvard University Press.

Notes

1. Nature – environment, the outdoors, the garden, green space, landscape, green world.
2. In the UK, an allotment is a plot of land rented usually from the local authority for growing fruit and vegetables. Allotment holders must adhere to contractual rules regarding land use, maintenance, and access in considering others' use of the shared space.
3. Musicians, artists and dancers may concur, finding ways other than the spoken/written word to express themselves.
4. I am referring to UK food systems here.
5. The content of vitamin C in kale for example, reduces by up to 5 per cent for each hour after cutting.

8 'THE CROP THAT RULED OUR LIVES'
Memories of Tobacco among Former Growers in Australasia

Andrew Russell

Introduction

Few plants have exerted, and continue to exert, so much influence in as many spheres of human life as tobacco (*Nicotiana tabacum*). It is roughly 500 years since European explorers and adventurers arriving in the so-called 'New World' enabled tobacco to escape its 'Old World' confines in the Americas. Here it had coexisted with Indigenous peoples for thousands of years. Since then, it has become global in its reach, exerting its power economically, politically, socially and biologically. Taking heed of the material culture and multispecies literatures that come together under the 'new materialities' paradigm, this chapter forms part of a larger project that acknowledges the agency of tobacco and its associated paraphernalia and the ways in which the ensuing assemblages produced have shaped the world in multiple and diverse ways. This chapter looks at tobacco as a crop and the objects that are necessary corollaries of its industrial-scale, plantation-style cultivation. I shall focus particularly on tobacco growing in Australia and Aotearoa/New Zealand. I shall do so in the context of the World Health Organization (WHO)'s Framework Convention on Tobacco Control (FCTC). Tobacco is the only plant, to the best of my knowledge, that has ever been the subject of an international health treaty. The FCTC went into force as a counterweight to tobacco and the power of the transnational tobacco corporations, which have turned what was once an exotic plant into an industrial commodity.

The FCTC has as one of its Articles the 'Provision of support for economically viable alternative activities', but examples of successful transitions from tobacco cultivation to alternatives are relatively few. Aotearoa/New Zealand and Australia are unusual in this respect in both being countries where the transition from tobacco to alternative activities has been effected (whether successfully or not being a matter for discussion). Tobacco growers' memories encompass nostalgia for a lost 'tobacco culture', although their stories about the way of life this entailed are bittersweet. Their narratives are rather different to those of contemporary growers, tending to focus more on their experiences of the growing practices they had to follow than the morality of tobacco growing per se. Their stories contribute to a greater understanding of tobacco cultivation elsewhere in current times, and offer fresh perspectives on reasons to be cautious about tobacco as a health risk. This is important because the narratives play out in the context of tobacco's current public health status as 'the world's greatest cause of preventable death' (Kohrman and Benson 2011, 329), a product with the prospect of causing 1 billion deaths in the twenty first century, 70–80 per cent in low and middle income countries where anthropologists such as myself have traditionally undertaken field research.

The Framework Convention on Tobacco Control: Article 17

The FCTC was the first ever treaty to be negotiated by the WHO. It has been one of the fastest growing and comprehensively endorsed treaties in the history of the United Nations. Within its 38 'Articles' are measures to reduce not only demand for but also the supply of tobacco. This chapter focuses on the supply side of the equation, and in particular those domains in which tobacco is a living and breathing plant, not the eviscerated, 'cured' substance its leaves are turned into for manufacturing purposes.

The FCTC's Article 17 is concerned with the promotion of 'economically viable alternatives for tobacco workers, growers and, as the case may be, individual sellers'. Progress in putting this Article into practice has been slow – the Convention Secretariat report published in 2016 describes it as one of 'the least implemented Articles of the Convention' with 'the scarcity of information on specific programmes and relevant research' being one highlighted cause (FCTC 2016, 2). The work that has been undertaken has tended to focus on growers rather than workers or sellers. I would argue that implementation problems in this sphere derive from the fact that tobacco cultivation takes place as part of a larger nexus of livelihood, way of life, and the phyto-power of tobacco. Tobacco forges strong and close connections with its human cultivators over time, just as it does with its human consumers. Substituting tobacco for another crop is rarely a straightforward exercise. For this reason, following Latour (2005, 2), I tend to use the term, 'tobacco assemblage', for all the actants and quasi-actants (Krarup and Blok 2011) that are glossed in the term 'tobacco culture' above. Such assemblages deserve careful analysis wherever and whenever they are found.

Contemporaneously, for example, we can look at the Indigenous connections between tobacco and people in lowland South America, where the plant's relationship with its human users and producers has been longest. 'We all have the same hearts, made from the same tobacco', one interlocutor explains to Londoño-Sulkin (2012, 102–03). Notes on the cultivation of tobacco appear in the Yanomamo folk tale recounted by Wilbert and Simoneau (1990), in which culture hero Tomï-riwë takes pity on the cravings of Haxo-riwë, and gives him a few tobacco seeds wrapped in a banana leaf and placed in a tube container.

> Take this [he says]. Go home, and then to your plantation and plant these in this way: clear a small area and burn the brush. Then take this package from the container and open it in your hand. When you sow, first blow on the seeds. Cover them with earth, and when they begin to grow, transplant the seedlings and cover them with leaves to protect them from the rays of the sun. Then you wait. When the leaves look large and beautiful, remove the nerves and place the leaves to dry on top of the hearth. After they have been dried like that they last for a long time. When you want to use them, place two in water to soak, pass them through the ashes on the hearth, and form a small roll, which you then place between your gum and your lower lip. When you're sucking tobacco like that you'll have the strength to work and won't feel so hungry.

In North America too, Winter reports, 'tobacco is so important in the [Eastern] woodlands that the Seneca, Cayuga, Onondaga, Oneida, and Mohawk…or Iroquois Nation, depend upon it for their very spiritual and cultural survival, as do their linguistic relatives the Huron, Neutral, and Khionontaternon (the Petun or Tobacco Nation)' (2000, 16). The Ojibwa, now found primarily around the northern Great Lakes region, are one of many other North American groups for whom tobacco is 'of supreme significance'. The Ojibwa cosmos has been described as a 'medley of voices by which different beings, in their several tongues, announce their presence, make themselves felt, and have effects. To carry on your life as an Ojibwa person you have to tune into these voices, and to listen and respond to what they are telling you' (Ingold 2013, 739). This is reminiscent of cosmological perspectivism, 'the conception, common to many peoples of the continent, according to which the world is inhabited by different sorts of subjects or persons, human and non-human, which apprehend reality from different points of view' (Viveiros de Castro 1998, 469) characteristic of Indigenous communities in South America where tobacco has for centuries been a 'master plant' (Russell and Rahman 2015). Tobacco used in shamanic proportions, I argue, has played a key role in establishing these multiple voices and perspectives (Russell 2019).

In terms of past records, we can look at the tobacco planters in eighteenth-century Virginia, for whom tobacco 'provided a medium within which the planter negotiated a public reputation, a sense of self-worth as an agricultural producer' (Breen 1985, 58). Such was the embodied nature of the crop, it came to be seen as 'an extension of self ... the highest praise one could bestow on an eighteenth-century planter was to call him "crop master", a public recognition of agricultural acumen' (Breen 1985, 61). In this way, tobacco's influence extended beyond the individual grower and permeated the entire society. One 1799 visitor to the Chesapeake region (where tobacco alone had ensured the survival of the original settler colonies and their subsequent growth) commented that tobacco 'was interwoven in the spirit of the time; and how nearly the history of the tobacco plant is allied to the chronology of an extensive and flourishing country' (Tatham 1800, 184). Finally, it is surely noteworthy how in the late 1760s Richard Henry Lee, in penning his unpublished essay 'The State of the Constitution of Virginea', devoted the bulk of its pages to the agricultural practices by which Virginians cultivated tobacco, with a view to informing those ignorant of its ways 'how much labour is required on a Virginean estate and how poor the produce' (cited in Breen 1985, 40).

With their incorporation into industrial-scale production methods, the relationship between tobacco as plant and people has also changed. The intense labour requirements of plantation tobacco cultivation were met by slavery in the USA and other parts of the Americas, and latterly child labour in countries like Malawi and Tanzania (Palitza 2011). These have been the main ways in which tobacco control advocates have tried to engineer disdain for the cultivation of tobacco on the global stage. Others have sought to understand how growers, like corporation executives, 'live with themselves' (Rosenblatt 1994). Pete Benson, for example, has sought to investigate how North Carolina tobacco farmers experience and act upon the 'moral ambiguities' of the tobacco business. As he puts it, 'by virtue of involvement in a problematized commodity chain, tobacco growers are forced to wrestle with society's changing attitudes about tobacco in ways that are often deeply personal and difficult' (Benson 2010, 501). He identified 'a stock script involving several patterned lines of defense' (Benson 2012, 136). Most farmers accepted the links between smoking and ill-health, but took a 'willed commodity fetishism' view that flattened the differences between tobacco compared to other products. The legality of tobacco was also a common trope. 'We are producing a legal product, just like any other business. Do I want my children to smoke? The answer is no. Do I think other people have a right to smoke? Yes,' one farmer told Benson (2012, 137). Other important elements in people's justification for tobacco growing was its economic contribution and its place in their heritage and identity. Religion also provided a distinctive mechanism or framework for justifying tobacco. 'Being a Christian, I want to be a righteous person in whatever way. But I grow tobacco. Does that make me un-Christian? Would Jesus grow tobacco? Would he vilify me for growing tobacco? I can't quit tobacco because I've got a family, and at my age' (Benson 2012, 140–1). Benson sees arguments such as these are 'core psychological defence mechanisms' deliberately crafted by the 'tobacco companies...for people associated with the industry' (Benson 2010, 500).

Benson's work is interesting, and unusual for its ethnographic focus on tobacco growing as a way of life (one other example is Kingsolver 2011). As Ferrell (2012, 127) puts it, 'the story of tobacco as it pertains to today's tobacco farmers is in many ways an untellable story because of the stigma now associated with the crop and, in many cases, those who grow it'. But what about communities where tobacco was formerly grown and where, due to rampant forces of geocapitalism or simply a change in the market, tobacco cultivation is no more. What are the stories that can be elicited through conversations with these former producers, and how do they differ from what contemporary growers might say?

Methods

In April/May 2019 I took advantage of an opportunity to work with former tobacco farmers in Aotearoa/New Zealand and Australia. Commercial cultivation of tobacco finished in 1995 in Aotearoa/New Zealand. Economic liberalisation, enabling the industry to switch to nations with cheaper labour costs and weaker labour laws, allied with the withdrawal of government subsidies, were largely to blame. Producers struggled on for longer in Australia, where the final legal crop was processed in 2006. I was interested in whether any remnants of the moral ambivalence Benson identified in North Carolina persisted and, if so, how these manifested. What were the differences between tobacco assemblages based on memory, heritage and legacy of former times, and contemporary assemblages in which tobacco production is a reason for defensive posturing? Can what we learn from the past inform how we handle the present?

I had originally become interested in tobacco heritage during a visit to Aotearoa/New Zealand in 2014. That research trip had been as a result of the country's ambitious plans to become tobacco-free (or rather, smoke-free) by 2025. I was surprised to find that efforts to banish the demon tobacco (as it had become) were not simply a matter of imagining a future world without tobacco, but of addressing a former world in which tobacco had played a significant part in the country's agricultural economy and society. As well as current tobacco-dependent lifestyles, there were former tobacco-dependent livelihoods to consider. In numerous communities in and around Motueka, a town in the north of South Island, evidence of the area's former tobacco heritage, both tangible and intangible, abounded. Motueka District Museum had created a tobacco exhibit in an effort to preserve this heritage. What happens when, in Freud's terms, a substance which was once normal, familiar and part of everyday life, becomes strange, alien and *unheimlich* yet remains an active presence in the lifeworld? In tobacco's heyday, during the 1960s and 1970s, temporary migrant labour from other parts of the country and even other South Pacific nations had been used to harvest the crop. Esteemed writers such as Keri Hulme had memories of taking part in the tobacco harvest in this way (see Russell 2019, 304–7). In attempting to eviscerate something as pervasive and embedded, not only within the individual body but also the social body and the body politic (Scheper-Hughes and Lock 1987), we need to be aware of and learn to cope with the spectral residues, what Fisher (2016) calls 'the eerie', marked by 'a failure of absence'.

In order to prepare for my greater engagement with tobacco as a plant rather than as a finished 'product', and with those who were once responsible for producing it, I decided to grow my own in the UK. I was acutely aware that despite my exposure to the paraphernalia of tobacco, and the plant in its manufactured, pre-combustible form, I had never knowingly seen a tobacco plant as a living entity. I attempted to rectify this omission with the purchase of five growing specimens of *N. tabacum* brought from Mel424 on eBay. Being involved in their seasonal progress in my back garden became a constant and welcome diversion to the rest of my research, which involved people, journals and books but also artworks, music, songs and other material objects. My tobacco plants afforded me fresh opportunities for engagement in a number of phenomenological, practical ways with the interests of growers and former growers. For my 2019 visit I intended to expand both my reach and depth by spending time with former growers not only in Aotearoa/New Zealand but also in Australia, and the former tobacco growing communities in the states of Victoria, Queensland and New South Wales that I was able to visit in the time I had available.

I was interested in what people had to say about their tobacco heritage and how they perceived, represented and managed both the crop and their identity as former tobacco growers in these places. Fieldwork took place over a three-week period. I visited former tobacco growers in three specific areas – in and around Myrtleford, in the Alpine zone of Victoria, Australia; in and around Motueka, in the north of South Island, Aotearoa/New Zealand; and in and around Inglewood and other places that had been important centres in southern Queensland and northern New South Wales, Australia. In each place I was keen to explore people's perceptions and actions with regard to their tobacco heritage. I was also keen to see the alternative livelihoods people had developed for themselves, and how successful these had been, in the wake of the disappearance of tobacco. I also realised a longer-term connection with outback NSW – in the 1960s my older brother had emigrated to Australia on the assisted passage scheme with a view to learning to fly, and had ended up in a small town called Wee Waa working as a marker for an aerial crop-spraying firm. My father, an airline pilot, had been able to finagle tickets for a family visit during Christmas 1968. At this time, aged 11, I experienced first-hand some of the sights that are to feature in the rest of this chapter, such as aerial crop spraying and the work of human 'markers' who stood along the flightpath of the crop spraying plane waving flags to ensure a decent coverage.

My subsequent engagement came through following tobacco as a now intangible heritage item. I first spent two days with the secretary of the Motueka Historical Society and the curator of the town's Municipal Museum, visiting the museum and former heritage sites and talking with four former growers in their homes. The first of my Australian heritage sites was Myrtleford, Victoria, where I had recruited possible interlocutors in advance via the internet. In Inglewood, Queensland, I was invited to stay with the President of the Inglewood Historical Association and was able to see local museum and other heritage sites, as well as investigating whether former growers were making a decent living and, if so, how. She also organised a wonderful barbecue to which former growers and other community members were invited and which turned into a spontaneous focus group. This was followed by visits in the company of the President, her husband and a research assistant to four former tobacco growers in their homes. We also visited former tobacco growing sites and places where former growers had established or were working in other industries.

I took notes on field observations, museum displays and informal conversations, and took plentiful photographs. Interviews and discussions were recorded with the permission of participants where it was convenient to do so, and were transcribed for further analysis. I collected old newspapers and other significant documents and took notes on museum displays. Further rich seams of information were provided by two published books in Aotearoa/New Zealand (Bastin 2018; O'Shea 1997) and a children's book (Robilliard 1967). I also collected three DVDs (two sold in their respective museums and one prepared by a former tobacco grower) and was alerted to a TV documentary recorded in New Zealand in 1964 and now available on YouTube[1]. A CD of songs recorded by former growers in Australia who had formed themselves into a small band added a further layer to the social history of tobacco growing in their communities.

The research plan was approved by the Durham University Anthropology Department's Ethics Committee. Participants were provided with an information sheet outlining the project. I subsequently revisited all these materials with a view to disseminating my findings to the people involved in their production as well as to a wider audience. All errors of fact or interpretation are my own.

Tobacco's traces

One looks at a landscape in the present and sees certain things. Someone who has known it in the past sees it differently, touched with the palimpsest of memory. Travelling around former tobacco growing communities in Australasia with those formerly involved in the business, the memories are clearly still strong, almost visceral. A car journey in Australia with one former grower was interspersed with statements like 'Tobacco grew all along the river flats here, as far as the eye could see. There would have been sharefarmers' cottages over there …'. As well as remembered buildings, the landscapes of former tobacco growing areas of Australia and Aotearoa/New Zealand are punctuated with the distinctive fingers of frequently abandoned and decaying kilns – predominantly of aluminium in Myrtleford and Motueka, of wood or concrete in the case of Inglewood. These are part of the still tangible heritage of the communities concerned. Whether, and if so how, to preserve this heritage is an issue that people are grappling with in each place. Depending on the use to which the land has subsequently been put, and the wishes of the owners, in some places these buildings are being destroyed. In other places, they are being repurposed. How this happened is partly dependent on what the people involved have decided, the parameters within which they are able to make these decisions, and their ingenuity and inventiveness in taking them forward. Off the Warangatta road leading into Myrtleford, Northeast Victoria, a former log tobacco kiln in built by the Pizzini brothers in 1957 has been transported from Eurobin and rebuilt with information inside about the history of tobacco in the area. Going out of the town in the other direction, Michelle and Ermanno Lupo have created a café in the style of a kiln made from recycled kiln materials (except for the internal beams).

The café opened in 2014. There are enticing homegrown vegetables in its exterior grounds. With real creative flair the shelves behind the bar are made from old seed boxes, the tables are old doors from the tobacco sheds, the supports are the old irrigation system pipes. There are photographs and other documents (e.g., music) under glass on the table tops. Some visitors recognise relatives under the glass; others from places such as Melbourne need to have the fact there is a tobacco history explained to them. Michelle would like to have done more with photographs/storyboards etc. on the wall, but the demands of the job were too much. The café has a one-bedroom apartment in the kiln loft part, which supplements the income from the two other holiday rentals. However, the demands of the hospitality industry took their toll (they have two small boys as well as other business interests) and they gave up operating the café themselves in 2018. They have now passed the running of it on to tenants who, in 2019 when I visited, were taking the strain of the constant demands of hospitality work.

Other tobacco-related buildings have been similarly repurposed or abandoned. On the road NW out of Myrtleford there is a large former TCV (Tobacco Cooperative of Victoria, subsequently Tobacco Company of Victoria) reception and warehouse. This was sold to the Alpine Valley Milling Group (AVMG) for AU$2.6 million in 2014 when shares in the TCV were distributed. The newspaper at the time describes the situation as the dissolution of 'the final remnants of the tobacco industry in Victoria and the North East' (Jenvey 2014). Former tobacco grower and TCV chairman Tony La Spina is reported as saying he and former growers were happy the whole process was about to finish up. However, the subsequent demise of the AVMG has left the warehouse as a bit of a white elephant on the edge of the town. In Aotearoa/New Zealand the former Rothmans depot, famous for its clock, has become a filling station and multi-functional retail centre. H. O. Wills's depot in Motueka has been repurposed as the town's Masonic Hall. The former Imperial Tobacco depot in the town has been less fortunate, turning into something of a junkyard (now abandoned).

In all the communities I visited, volunteers interested in the past have made efforts to memorialise this heritage through the creation of tobacco exhibits – in the district museum in Motueka, through the creation of the Australian Tobacco Museum in Inglewood, and a heritage centre in Texas, southern Queensland[2]. There are longstanding plans in Motueka to establish a Heritage Centre to complement or perhaps even replace its museum, which is located in a former school house and is bulging at the seams. Members of the Heritage Centre committee have collected a large number of machines and implements from the tobacco era they would like to display, although the museum curator is concerned that committee members may be more like 'collectors' rather than 'displayers' of memorabilia. After all, 'how many transistor radios do you need to show what it was like to live in a bach?' (the housing quarters for migrant labourers during the tobacco era) she asked. Very often the heritage is recorded on film, in photographs or video. One of the most exciting aspects of my visit to Motueka was catching up with a tobacco picking machine that was decaying in the corner of a former tobacco-growing paddock. This machine featured in the Motueka episode of the 1964 documentary series 'These New Zealanders' devised by Selwyn Toogood.

Hidden voices

Films like Toogood's are important because some of the intangible heritage – the oral traditions, performing arts, local knowledge, and traditional skills – are disappearing as people move away, move on or die. This process might be accentuated because the crop is tobacco. While not frequently expressed, some people clearly felt a measure of embarrassment at having been involved in what they knew at the time and look back on now as such a dirty crop, literally and metaphorically. 'I hated growing tobacco – it had such a bad name in the 1990s. You'd go away somewhere, and people would ask where you're from – you only had to tell people Motueka for them to assume you were in the tobacco business,' remembered one woman. For Patsy O'Shea, introducing a talk by the former manager of the Rothman's tobacco factory in Motueka, 'in this town it was an essential product, an essential part of our economy for so long that we need to stay interested in it, even if other people think it is a topic non grata to speak about'[3].

There were shades of the pride in tobacco's past similar to those expressed by their North Carolina counterparts today. Many interlocutors appeared nostalgic for the prosperity tobacco had brought. One respondent reported after tobacco cultivation had ended in New Zealand he went on holiday to Australia, where tobacco growing was still continuing, and had been so moved by the sight he needed to get out of the car just to feel the leaf again!

At a more fundamental level, tobacco was still an integral part of a grower's being. 'It's more than just a crop, it's a way of life, an identity', said a former grower in Myrtleford, who went on to say tobacco was such an important part of his life that losing it was like losing a member of the family. According to him, while the State is against tobacco on public health grounds, he and friends used to smoke it straight out of the kilns, and he never found it to be an addictive substance. It was also the source of human conviviality. In Myrtleford a long shed on the side of three kilns that was used for bale storage had originally come from the Snowy River hydro scheme and was remembered for its dances – Italian, Croatian, all could use it. There were also barbecue socials.

Others, however, remembered tobacco growing only with distaste bordering on abhorrence. 'The day we stopped growing tobacco was the best day of my life' said one former grower at the Inglewood barbecue. 'The same goes for me' interjected another guest. One man reported still having nightmares once or twice a week over 'Have I turned that irrigation off?' 'Have I started the bulk curers?' Bastin's memoir of the Motueka area shows her to be no great fan of tobacco. 'I still remember those harvesting seasons with a surprising amount of pleasure. Which only goes to show how the glimmering mist of nostalgia can distort memory', she writes (Bastin 2018, 107). One problem she identifies is that, 'because a lot of tobacco work was done by family members, the returns most growers got for their tobacco in the sixties did not reflect the actual costs of producing it' (Bastin 2018, 61). She goes on, 'like other growers, my father had some good years growing tobacco, and he had some particularly bad ones too. Whichever was the case, all our lives were ruled by that omnipotent despot of a crop' (Bastin 2018, 63).

Another reason people's tobacco stories are less frequently heard, in Australia at least, is because many people in the business were

Italian or of Italian extraction. Many came to Australia with little money and the women in particular lacked English language abilities and varied in access to tuition and the ability to pick it up. Some of the interviews I conducted (because I lack facility in Italian) took place in heavily accented English, some seventy years after people first migrated from their homeland. Much twentieth century tobacco cultivation, particularly around Myrtleford but also Inglewood, was taken up by Italian migrants. They often operated as sharefarmers to an existing (and usually non-Italian) family initially, before buying their own land. One family traced their migration back to a grandfather who first came from Italy in 1921 but subsequently returned to his homeland only to sell up everything there and return to Australia in the 1930s. However, his grandmother and an uncle were delayed from returning by the war – the uncle disappeared during an allied bombing raid in Italy before the end of the war and hence never made it.

Migration could either be permanent, of families from another country, or temporary – the seasonal migration of people from one part of the country or region to another. The first people to grown tobacco commercially in Australia were the Chinese from the 1800s onwards, their names and memories surviving in a few grainy and frequently fading photographs. In New Zealand there was some resistance to the idea of importing cheap 'coolie' labour to produce tobacco, since this was seen as going against the principles of fair reward for honest labour on which the state had been founded and, it was feared, would undermine the wages of European New Zealanders (O'Shea 1997, 12). In other places groups were displaced to make way for settler communities and the tobacco they grew. The aboriginal groups around Inglewood shire in Queensland were the Kambuwal to the east and the Bigambul to the west. They were at the epicentre of the 'death pudding' (flour mixed with strychnine and arsenic) that was commonly used by settlers in Queensland to poison unsuspecting aboriginal groups (Elder 2003, 144). I was told of a massacre that took place sometime in the first half of the nineteenth century, when shepherds reported aborigines were amassing near a settler station for purposes that were interpreted as hostile by the settlers. In order to pre-empt any trouble, the owner of the station ordered the poisoning of their waterhole with arsenic. They all died. In fact, this is a rather improbable story, at least the waterhole part. Given the dependence everyone has on waterholes in the outback, it is more likely that the usual 'death pudding' was used. 'No-one talks about that very much', said the narrator of this story.

All these are hidden voices, obscured for various reasons in and through the practice of tobacco cultivation. However, other kinds of invisibility are exposed not through talking about their perspectives, but by reflecting on their practical input into the life of the plant itself and the methodical, scientific approach and experimentation that took place in some cases. It is to some aspects of these phyto-technologies and techniques that I shall now turn.

Technologies and techniques

Everyone remembered the tobacco year as being hard work. It began with sowing the sterilised seedbeds in September. The dust-like quality of the seeds meant they could be mixed with irrigation water and sprayed onto the seed beds by watering can or irrigation pipe. In Myrtleford the harvesting had to take place in March/April to avoid a risk of frosts. People in Myrtleford reported tobacco cultivation had become easier as time went on, with increasing mechanisation. In the past, tobacco was graded into around 50 different grades. Latterly it was just a question of 'good bale, bad bale'. The government used to charge AU$35,000 excise per bale.

The agricultural cycle dictated by tobacco was extreme indeed. In earlier days people would work through the paddocks cutting off leaves that were ready and gathering them under their arms in a gunny sack. Looking at a picture of stacks at the end of rows waiting to be collected for processing, one former grower commented there was a politics of picking in that the people nearest the end of the rows had less distance to walk back and forth than those further in, so more established pickers tended to get these areas. As time went on harvesting became more automated, as it does for many crops as economies of scale kick in. Aotearoa/New Zealand probably had the march on Australia in terms of harvesting technology. The most straightforward picker involved a raised machine running between rows. As shown on the Selwyn Toogood documentary mentioned above, pickers sat on the bottom picking off the leaves into sacks along the centre of the machine. A more sophisticated double-decker machine came from the USA. The picked leaves went on a conveyor belt up to the top deck where a team of four women waited to sort the leaves into sacks which could then be loaded onto trailers. In the most sophisticated cases, trailers were actually attached to the machine, lifted to the higher level by hydraulics. A version of this machine had been tried in Australia but had never worked successfully.

Perhaps the Australians should have taken a leaf out of the citizen science experiment of the New Zealand grower, who explained:

> you see those two little wheels there ... it was like a scissors handle, and I put springs on them to shut them, and you just rolled the leaf into them and away it went. But by the time we'd picked our first kiln, half the clips wouldn't hold any leaves, and I thought ... and we'd just started harvest, and I thought 'this isn't going to be any good'. Somebody said 'you need a rubber band – it'll fix anything'. I thought 'rubber band', and I put three on each clip and they did the whole year. But the gum, the juice on the leaf when it sets, the hair on your arm would be standing up as sticky as blazes. And the coil on the spring it got in between and it wouldn't close. But I got the rubber bands on and away it went.

The harvested bundles were taken to the sorting shed, where they had to be unpacked and prepared for curing. In earlier days, leaves were tied together and arranged over sticks by hand. Then a ramp was introduced with a giant sewing machine at one end to automate the process. Everything was then ready for curing, which took place on the premises. Loading the kiln consisted of men climbing up and arranging these sticks from the top beams in the kiln to the lower beams – according to one woman, only her father was permitted to do it because of the fire risk. Then the furnace was stoked with firewood, with pipes leading from it along the floor, for the curing process to start. It took five days and the temperature had to be maintained at a steady 35–40°C. There was always the risk of fire in the kiln from leaves or sticks that became dislodged and fell onto the hot pipes. Some kilns had rolls of chicken wire mesh between leaf and pipe in an attempt to prevent this, but this was not a failsafe preventive.

The bulk loading downdraft kiln may have been an attempt to reduce the fire risk. Rather than sticks, leaves were gathered and pushed into bulk loading bins using pronged loaders, a particularly strenuous job. Pipework from the furnace was channelled up to the top of the kiln and heat permeated downwards through the bins. The process could not be rushed – to do so meant the leaves maintained too high a starch content rather than breaking down into sugars, and this made for an unpalatable smoking leaf, according to one Motueka grower.

The curing was completed, somewhat paradoxically in a process intended to dry out the leaf, by a period when a steamer was used. This was necessary to make the leaves sufficiently pliable to work. The steamer took the moisture content of the leaf up to 20 per cent or so, but there was a redrying tunnel at the buying shed to bring the moisture content down to around 12 per cent, which was necessary for long-term storage. Cured leaves were then removed from the kiln and cut from their sticks (where used). Before its simplification into 'good bales' and 'bad bales' there then came the long winter process of grading (see, for example, Bastin 2018, 115–18).

Many growers took a scientific approach to their crop, testing different varieties, chemicals and techniques against each other. 'Dad was big on it,' said one former tobacco farmer, now growing lucerne (*Medicago sativa*) in retirement.

> We always did experiments for CSIRO, the local department up here, and still to this day, even with my lucerne, I do experiments, try different things. We always had a control, and that was one thing I was arguing about with my cousin, down there, he's got lucerne too. He got this [new] stuff ... and I said to him 'Did you put it over everything' and he said 'Yeah', so I said 'Well how do you know if there's any difference?' When I did it here I left there a bay, and I picked out a certain area I knew was pretty well the same so that I could see if that bay was any different from this other one. And I had the bloke here that we did the experiment for, they call it Tiamagum, it's supposed to be – they wouldn't tell you what was in it, it was supposed to be some magic bloody formula, and all it was supposed to do was to get your bacteria working better in the soil and that. And this bloke come to check it out and I said to him have a look here and dug and he said 'yeah, that doesn't look too bad, your humus is good' and all the rest of

it, and I said 'have a look over here', and he said 'Oh, this looks even better'. And I said 'yeah, well that's the one that got none on it'. And he said 'Ooh, I've put my foot in it haven't I?' And I said 'You don't know unless you do that'. And that's something we always did.

Now units such as the CSIRO are closing or have shifted to different crops and priorities. One former local research station in Inglewood has been extended and converted into a family home. In Motueka the new priority is hops and hop varieties. Interestingly enough, both the Motueka and Inglewood Museums had recently acquired extensive folders of scientific reports from their former local research stations and were struggling to know what to do with them. There would be a good project here for a student interested in agricultural history either exploring an individual research station's history or comparing and contrasting the outputs of the two research stations in their time.

These kinds of practical interests and conversations are somewhat different to the more theoretical focus of many anthropologists with an interest in Indigenous life. Climate, labour, machinery, soils, manufacturing and crop diseases (Russell 2019, 308–10) are popular conversation topics, rather than the politics, citizenship, affect and corporate tobacco elements that are the more sociological and psychological aspects of Benson's analysis (2012).

Crop hazards and their control

Tobacco was regarded as vulnerable to all sorts of environmental hazards. Lots of people in Inglewood remembered the big flood of 1956 and another in 2011. 'It was a disaster for the tobacco' said one lady. Hailstorms were also a risk when the crop was ripening. Frosts in colder places such as Myrtleford and Motueka before harvesting could take place were devastating.

Another primary theme in the story-telling about tobacco in these regions, was how the more biotic problems were tackled, the chemicals – 'poisons' many people called them – used in its cultivation. Products abounded to sterilise the seedbed soil, to fertilise the ground, and to control the various plant and animal pests which, although nicotine had evolved as a powerful insecticide in its own right, were quick to capitalise on a farmers' lack of protection in this regard.

The steriliser of choice was methyl bromide, 'easy to handle and quick to disperse and do its job', one former NZ grower told us. Here it had to be administered under plastic sheeting to release it in its gaseous form in a controlled manner, or else 'when you puncture your tin to shove it in if something happens and it doesn't seal properly, you drop the damn thing on the ground and get the Hades out of it! Hold your breath and go and after about an hour you can go back and sort it out'. One former grower in southern Queensland remembered a dog who had sniffed it and died.

Like methyl bromide, insecticides such as Dieldrin, Aldrin, Endosulfan, are all products that are now withdrawn, banned or very tightly controlled. Their administration also challenged environmental standards. People in southern Queensland remembered the aerial crop-spraying.

> When I first got married, we were in this little cottage ... and the planes would come in where the crops had got a bit too big to walk in with a knapsack spray. I could look down the rows and the planes would be coming in the tobacco swooshing around and it would swoosh up over my house I you could hear the drops falling on our roof. Then it would rain and the drops would go into your water tank and you'd fill up and make a cup of tea ... we didn't consider the possibility it was dangerous to do that.

Another woman could remember her father asking her to read him the instructions on how to mix the chemical poisons.

> I was only a child – I could read it but I didn't understand ... 'You just mix it with your hand ...', 'and just put a bit more ...', and there was no real measurements either. You could say 'Ah yeah, a bit more won't hurt', and they'd stir it up and put it in the knapsack spray. I was nursing at the time and you'd get them into the hospital sick as dogs, vomiting, so sick. And we'd fix them up sort of put drips in and drain them out a bit, and they'd leave and go back and do it again. It was Dieldrin and Aldrin, and that's just appalling stuff. It's all banned now.

For one man, 'it was a major excitement, a distraction at the school in Whetstone watching the very accurate pilots flying down right beside a line of telegraph posts eight feet above the ground'. 'I remember seeing the spraying going everywhere and it didn't occur to me that that stuff was [causing] damage'. The spraying had hidden consequences for the quality of the end product, the cigarette. According to one former grower, 'nicotine was probably the safest thing in a cigarette, because there was no withholding people, we'd finished spraying bloody DDT or Dieldrin, and we'd pick it the next day, and it all went into the cigarettes'.

I thought it interesting that of all the shifts in agricultural practice, and out of agriculture altogether, that had occurred as a result of the end of tobacco cultivation in Australia and New Zealand, this latter grower, one of the last in southern Queensland, had made perhaps the most radical shift, in his case into organic farming. This was not an easy path to take (one has to keep one's ground crop-free for three years to enable any residual chemicals to leach out of the soil). This farmer has not only followed the organic method, but has become secretary to an organic accreditation committee. He described it as like some kind of atonement, if not for his heavily tobacco-dependent livelihood, then for the agricultural practices that made it possible.

The demise of tobacco

Although some (including industry sources) attribute the end of tobacco growing to the public health lobby and a consequent decline in sales, most are aware that the main reason for its demise was that companies could source tobacco cheaper in other countries. However, there was resentment towards the government in some places. 'If the government was fair dinkum, they'd ban it completely' said one Myrtleford man. The end, when it came, was swift. In Myrtleford the general opinion I picked up was that the demise of tobacco dealt a body blow from which farmers are still recovering. 'It was traumatic for all of us' said one man. Some farmers went into sorghum and lucerne for hay initially, then went on to blueberries and cattle. Others tried zucchini and pumpkin, but they were generally unsuccessful ventures. Although global markets had also caused the end of tobacco growing, farmers felt vulnerable to price fluctuations in a way they hadn't been in the case of tobacco (apart from in times of crop failure). One former tobacco grower in Myrtleford who had shifted into blueberries complained of a glut on the national market in 2019 so he got less than his production costs for them. This was something he attributed to Indian farms on the Gold Coast producing massive acreages of blueberries flooding the market. Some were nostalgic for the quota system, which had worked well in its day. One farmer in Myrtleford reported how farmers in Israel were organised according to quotas. However even if someone signed up to join such a quota system, it was not automatic that you would be awarded one.

Reflections

In a short fieldwork period such as this it is impossible to gain a comprehensive picture of the local histories of tobacco, the people who farmed it and the contexts in which they did so. After the ensuing Covid-19 pandemic there would be scope for social science students from regional, national or international universities to come and engage in much longer periods of fieldwork, during which time they could travel round the area visiting former growers and achieving a much more systematic mapping of the tobacco landscape. Given the origins of many of the former tobacco growers in Australia at least, it would be good if this student were fluent in Italian. There are also other communities to be visited in order to extend the reach of the tobacco story – Mareeba in North Queensland, for example, the Glasshouse Mountains (also Queensland), Roma (NSW), Western Australia (Baker 1960), and North Island New Zealand. Even on my last day I was told about someone who had grown a crop of tobacco near Tenterfield, Queensland. They had built their own kiln, but had not pursued the venture beyond one season.

My research differed from Benson's because I was working with former rather than current tobacco farmers. I hadn't bargained on the ramifications of this in terms of the diminution of Benson's 'core psychological defence mechanisms' (assuming they once existed), once the corporate circus that is the tobacco industry had left town (or region, or community). Except where old narratives of pride or embarrassment in tobacco appeared to have been internalised either individually or at the group/communal level, most people had nothing to hide (if they ever felt the need to do so). It was maybe this that enabled the deeply scientific approach many took to cultivating this plant to emerge. This should have come as no surprise. After all, Rival (2014, 226), reflecting on fieldwork in Amazonia, writes 'while I often struggled to get someone to help me collect stories about shamanism or to transcribe myths and chants, I had no difficulty in finding people willing to show me how to prepare a banana plantation, or to explain the specific uses of a particular plant'. Such a willingness would doubtless have extended to discussions about tobacco, as it did for my interlocutors in Australasia.

Focusing on practical measures in this way also meant there were sometimes candid revelations about agricultural practices with regard to tobacco cultivation that would probably not have occurred in a more defensive, guarded research context. Focusing on tobacco's heritage rather than contemporary practices (except where these were part of the alternatives to tobacco) meant that people were much more likely to focus on technical, material issues such as climate, labour, machinery, soils, manufacturing and crop diseases (Russell 2019, 308–10) and less on the politics, citizenship, affect and tobacco corporations that are the keystones of Benson's analysis (2012). Conversations about tobacco tended to morph into discussions about techniques and technologies (because that was the gist of tobacco's social life on a daily basis for many participants). From these candid discussions came information about a particularly important element, namely the chemicals that became the mainstay of industrial-scale tobacco cultivation.

I have since read up more on the health effects of short and long-term exposure to pesticide use, particularly the risks of neurodegenerative diseases. Parkinson's disease (Rösler et al. 2018), dementia (Yan et al. 2016), 'cognitive, behavioural and psychomotor dysfunction' (Zaganas et al. 2013, 3), increased cancer risks and the risk of teratogenic effects are among the dangers. Dieldrin in particular has been linked to the earlier onset of Parkinson's disease such that 'a person who is destined to get Parkinson's because of genetics or other factors at age 80 might develop symptoms when they're 65 or 70 if they have been exposed to pesticides' (American Chemical Society 2006). Going back to my own early years experiences in 1968, I wonder also about the fate of my father. He went out on many more evening crop spraying expeditions in 1968 than me, was diagnosed with Parkinson's Disease in 1973 and died of a brain tumour in 1981. However, the environmental hazards to which flight crews are exposed (e.g., Dreger et al. 2020; Meier et al. 2020) mean that other, occupationally derived health factors might have been at play in his case. Meanwhile my brother, who worked with pesticides for much of his life in Australia, died in February 2020 from prostate cancer following periods of serious memory loss and other ill-defined neuropathies. Of course, the problem with establishing causal or even predisposing linkages is the large number of intervening variables likely to build up over a lifetime, such as the high-altitude cosmic radiation to which flight crews are exposed and a variety of so-called 'lifestyle' factors.

Conclusion

I have argued that, for former farmers in Australia and Aotearoa/New Zealand, moral ambivalence about their crop exists less in the nature of the crop itself, since it was always legal and frequently remunerative, and more in the chemicals (or 'poisons', as they are often called) used to manage its cultivation by spraying and other modes of administration. I have highlighted what some of these were, and how marginalised the voices remembering them have frequently become. They demonstrate that working with former rather than current tobacco growers who have 'nothing to hide' in their reminiscences and story-telling about the crop, gives a different picture of tobacco production to studies involving current growers. In particular, the heavy use of chemicals in tobacco cultivation – such as steriliser, fertiliser, pesticide and herbicide – raises questions that are seldom voiced about a product that comes with a health warning but never with an ingredients list.

As time passes, and due to the fact that the health legacy of these different noxious substances is a matter of risk and probability rather than definitive causation, concerns over their long-term legacy are only infrequently and speculatively voiced. A restitution narrative is developing whereby the toxic legacy of these poisons is being replaced by a new ethos of organic farming methods free of chemical inputs. Retrospective research methods such as delving into oral histories in their ethnographic context, as reported here, bring fresh perspectives to bear on the present when dealing with a morally ambivalent subject like tobacco.

References

American Chemical Society, 2006. 'Pesticide Exposure Could Increase Risk of Early Onset of Parkinson's Disease', *ScienceDaily*, 15 September 2006.

Baker, A. E., 1960. 'Tobacco production in Western Australia', *Journal of the Department of Agriculture, Western Australia*, Series 4 1(11), Article 5.

Bastin, S. H., 2018. *Nicotine Flavoured Scones for Smoko – Stories of Tobacco Growing in the Motueka Region*. Christchurch: Printing.com.

Benson, P., 2010. 'Tobacco talk: reflections on corporate power and the legal framing of consumption', *Medical Anthropology Quarterly* 24(4), 500–21.

Benson, P., 2012. *Tobacco Capitalism: Growers, Migrant Workers, and the Changing Face of a Global Industry*. Princeton, NJ: Princeton University Press.

Breen, T. H., 1985. *Tobacco Culture: The Mentality of the Great Tidewater Planters on the Eve of Revolution*. Princeton, NJ: Princeton University Press.

Dreger, S., Wollschläger, D., Schafft, T., Hammer, G. P., Blettner, M. and Zeeb, H., 2020. 'Cohort study of occupational cosmic radiation dose and cancer mortality in German aircrew, 1960–2014', *Occupational and Environmental Medicine* 77, 285–91.

Elder, B., 2003. *Blood on the Wattle: Massacres and Maltreatment of Aboriginal Australians since 1788*. Sydney: New Holland.

FCTC, 2016. 'Economically sustainable alternatives to tobacco growing (in relation to Articles 17 and 18 of the WHO FCTC'. Report by the Convention Secretariat FCTC/COP/7/12.

Ingold, T., 2013. 'Dreaming of dragons: On the imagination of real life', *Journal of the Royal Anthropological Institute* 19(4), 734–52.

Ferrell, A. K., 2012. '"It's really hard to tell the true story of tobacco": Stigma, tellability, and reflexive scholarship', *Journal of Folklore Research* 49(2), 127–52.

Fisher, M., 2016. *The Weird and the Eerie*. London: Repeater Books.

Jenvey, J., 2014. 'Industry butts out', *Myrtleford Times*, 17 December 2014.

Kingsolver, A. E., 2011. *Tobacco Town Futures: Global Encounters in Rural Kentucky*. Long Grove: Waveland Press.

Kohrman, M. and Benson, P., 2011. 'Tobacco', *Annual Review of Anthropology* 40, 329–44.

Krarup, T. M. and Blok, A., 2011. 'Unfolding the social: quasi-actants, virtual theory, and the new empiricism of Bruno Latour', *Sociological Review* 59(1), 42–63.

Latour, B., 2005. *Reassembling the Social: An Introduction to Actor-Network-Theory*. Oxford: Oxford University Press.

Londoño-Sulkin, C., 2012. *People of Substance: An Ethnography of Morality in the Colombian Amazon*. Toronto: University of Toronto Press.

Meier, M. M. et al., 2020. 'Radiation in the atmosphere – A hazard to aviation safety?', *Atmosphere*, 11, 1358, doi: 10.3390/atmos11121358.

O'Shea, P., 1997. *The Golden Harvest*. Christchurch: Hazard Press.

Palitza, K., 2011. 'Child labour: the tobacco industry's smoking gun', *The Guardian* 14 September 2011.

Rival, L., 2014. 'Encountering nature through fieldwork: expert knowledge, modes of reasoning, and local creativity', *Journal of the Royal Anthropological Institute* 20(2), 218–36.

Robilliard, R., 1967. *Kay of the Tobacco Farm*. Wellington: Kea Press.

Rosenblatt, R., 1994. 'How do tobacco executives live with themselves?', *New York Times Magazine*, 20 March 1994, 22–34.

Rösler, T. W., et al. 2018. 'K-variant BCHE and pesticide exposure: Gene-environment interactions in a case-control study of Parkinson's disease in Egypt', *Scientific Reports* 8:16525, doi: 10.1038/s41598-018-35003-4.

Russell, A., 2019. *Anthropology of Tobacco*. London: Routledge.

Russell, A. and Rahman, E. (eds), 2015. *The Master Plant: Tobacco in Lowland South America*. London: Bloomsbury.

Scheper-Hughes, N. and Lock, M., 1987. 'The Mindful Body: A Prolegomenon to Future Work in Medical Anthropology', *Medical Anthropology Quarterly N.S.* 1(1), 6–41.

Tatham, W., 1800. *An Historical and Practical Essay on the Culture and Commerce of Tobacco*. London: Vernor and Hood.

Viveiros de Castro, E., 1998. 'Cosmological deixis and Amerindian perspectivism', *Journal of the Royal Anthropological Institute* 4(3), 469–88.

Wilbert, J. and Simoneau, K. (eds), 1990. *Folk Literature of the Yanomami Indians*. Los Angeles, CA: UCLA Latin American Studies Center.

Winter, J. C., 2000. 'Traditional uses of tobacco by Native Americans', in J. C. Winter (ed.), *Tobacco Use by Native North Americans: Sacred Smoke and Silent Killer*. Norman, OK: University of Oklahoma Press, pp. 9–58.

Yan, D., Zhang, Y., Liu, L., and Yan, H., 2016. 'Pesticide exposure and risk of Alzheimer's disease: a systematic review and meta-analysis', *Scientific Reports* 6, 32222, doi: 10.1038/srep32222.

Zaganas, I., Kapetanaki, S., Mastorodemos, V., Kanavouras, K., Colosio, C., Wilks, M. F. and Tsatsakis, A. M., 2013, 'Linking pesticide exposure and dementia: What is the evidence?' *Toxicology* 307, 3–11.

Notes

1. These New Zealanders No. 5 Motueka (1964). *https://www.youtube.com/watch?v=4OGuF-O3NNw*.
2. Myrtleford also has a local history museum, but I was unable to visit it due to its rather constrained opening hours.
3. Patsy O'Shea, introduction to Geoff Tillson, 'Rothmans Tobacco Factory Motueka'. A talk given to the Motueka and Districts Historical Association, October 2012. Transcribed MS in the possession of the Association.

INDEX

A

absorption 3, 11
   see also self-absorption
affect 2, 5, 10–11, 14, 20, 21, 42, 50–2
   passim, 60, 105, 112, 123, 129,
   205, 209
agency xx, 12, 23, 37, 51, 53, 69, 113–15
   passim, 129, 135, 148–9, 152,
   169, 183, 189
   see also non-agential
Agency, Environment 72
agriculture 39, 41, 43–4, 48, 50, 67,
   207
air 1, 4, 14, 25, 94–5, 109, 116, 118,
   127–9 *passim*, 145, 167
airways 41
Aldrin 206
Aluna 15, 84, 87–90 *passim*, 92, 95–7
   passim, 101–2, 104, 106
Amazon 11, 12, 61, 63, 83, 181
   Amazonia 110, 209
   Amazonian 88, 102, 103
animal/s 9, 17, 19–23 *passim*, 37–41
   passim, 44–5, 51–3 *passim*, 65,
   72, 73, 83, 87, 90, 95–6, 111,
   113, 116, 120–1, 125, 127, 137–43
   passim, 145, 150, 172, 175
   animal bodies 136
   animal embryos 137
   animal pests 206
      also see herbivores
   animals, splitness of 138
   giant animals 94
   human animals 19–20, 136–7, 150
   non-human animals 8, 17, 40, 163
Anthropocene, the 8, 22, 57, 67, 135,
   154

Anthropocentrism 145–6
   Anthropomorphic 146
   Anthropomorphises 19, 145
antimicrobial 148
Aotearoa (New Zealand) 12, 189,
   194–8 *passim*, 202, 210
archaeology 37, 40–2, 50–3 *passim*
   archaeobotanical 43, 57
   archaeobotany 41
   environmental archaeology 51
   zooarchaeology 77
Aristotle 20–1
Australia 12, 28, 63, 64, 190, 194–7
   passim, 200–3 *passim*, 207–8, 210
   Australian 64–6 *passim*, 196, 199,
   203

B

Baka, the 18
balance 7, 62, 66, 92, 94, 105, 171
Bantu 126, 129
Barad, Karen 3, 6, 7, 23–4, 26, 29
   Baradian queer 5
bark 66, 121, 125
Bateson, Gregory 1, 5–7 *passim*, 24–5,
   115, 120, 129
Bateson, Nora 25
beauty 37–40 *passim*, 51–2
becoming 2–4 *passim*, 14, 18, 28, 44,
   47–8, 51, 112, 115–17 *passim*,
   122–5 *passim*, 129, 138, 149
   becoming-plant 116, 125
   co-becoming 28
   deleuzoguattarien becomings 125
Bennett, Jane 109, 113, 114, 146, 170
biocultural diversity 63, 73

biodiversity 63, 69, 72, 73
blood 87, 90, 96, 153
　blood sugar levels 66
body xviii, 3, 11, 18, 19, 41, 42, 52, 62, 65, 86, 97, 101,112, 113, 124–6 *passim*, 142, 147, 151–2, 154, 169, 179, 195, 207
bodies 2–6 *passim*, 9–10, 13, 17, 20, 25, 87, 95, 12–15, 128–9, 136, 138, 150, 153, 170, 172, 175, 177
　body social 195
　body politic 195
　body type 3
　body without organs (BwO) 124, 140–1
　mind body 42, 52
　Mother Earth's body 95–6, 105
　object-body 136
botany 112, 149
botanical 11, 13, 16, 17, 57, 67, 99, 111, 129, 142
　botanical object 111
　ethnobotanical 5
　see also archaeobotanical
bush tucker 66

C

Cameroon 110–11, 116–17, 123, 129
cannabis (*Cannabis sativa* L.) 42, 57
cigarettes 11, 207
　see also tobacco
co-dependency 47
co-evolution 38, 47
colonial 28, 64, 68
　colonialism 60, 62, 70
　see also decolonise; decolonising
communication 19, 45, 69, 96, 139, 143, 148
　communicate; communicated; communicating, communicative 6, 8, 11, 12, 14, 19, 20, 24, 89, 94, 96, 100, 102, 104, 110, 172
　see also plant communication

composing with plants 20, 109, 117, 129
control 28, 39, 48, 52, 65, 109, 110, 129, 193, 204–6 *passim*
　control over nature 52, 65
　controlled trials 60
　reproductive control 62
　see also Framework Convention on Tobacco Control (FCTC)
Cosmic Pillar 86
cosmology 15, 86, 112, 118, 128
Covid-19 22, 70, 161, 163, 165, 167, 180, 182, 208
crop 11, 12, 39, 40, 43–9 *passim*, 51, 52, 94, 101, 170, 190–6 *passim*, 199, 200, 202, 204–9 *passim*
　crop dispersal 42
　crop-spraying 207
　crop, life cycle of 44, 47
cultivate 39, 43, 47, 48, 67, 71, 127, 192
cycle 90, 102, 106, 203
　growth cycle 40, 90
　life cycle 44, 47, 171
　water cycle 85, 94
　see also crop life cycle; recycled

D

data 25, 61, 64, 71, 169, 177
　big data 25
　sensory data 169
　warm data 25
database 61, 64
decolonise; decolonising 27–8
deforestation 51, 110
Deleuze, Gilles 110–15 *passim*, 118, 124, 129, 138, 140
　deleuzoguattarien becomings 125
dependency; dependence; dependencies 2, 10, 26, 49, 57, 73, 92, 201
　co-dependency 47
　see also interdependencies
desire 38–40 *passim*, 42, 51, 52, 111
diabetes 66, 76, 78
Dieldrin 206, 207, 209

diet 9, 41, 66, 175
  hunter-gatherer diets 34
discerning 109, 116, 129
disease 49, 57, 65, 126
  crop disease 170, 205, 209
  Parkinson's disease 209–10
  see also sickness
dispersal 72
  mechanisms 48
  of crops 42
  of pollen 38
  of seeds 48
Doctrine of Signatures 62, 66
domestication 9, 38, 39, 43, 47
dominance 7, 26, 170
drugs 58, 59, 68
  chemotherapy drugs 58
  pharmaceutical drugs 58, 71

E

Earth 63, 67, 83, 85, 89–91 *passim*, 95, 96, 105, 180
  'bald earth' 90, 96
  earth 37, 86, 87, 91, 94, 124, 125, 182, 191
  earthy 153
  Earth-Sun axis 137
  earthquakes 90, 105
edges 4, 26
edibility 9, 137, 146
  Edibility Approach, the 20, 22, 135, 141, 146–7, 150, 171
embodiment 16, 37, 41, 42, 44, 86, 178–9
embryology 21, 135–6, 141
emplantment 42
endosulfan 206
enslaved 51, 52, 127
  enslavement 21
entanglements 37–8, 41–3 *passim*, 47–9, 51–3 *passim*, 116, 127, 154, 182
environment 53, 69, 148–9, 154
  Environment Agency, the 72

environmental archaeology 51
environmental pollutants 72
giving environment, the 9–10, 15
escape, the garden as an 165, 166, 168
ethnobiology 63–4
ethnobotanical 5
ethnobotanist 62, 65, 69

F

Fabric of Life 86, 89
farming 8, 44–5 *passim*, 47–8 *passim*
  organic farming 207, 210
  see also domestication
feel; feeling; feelings 4, 50, 89, 101, 105, 142, 151, 164–6 *passim*, 167, 169, 176, 180, 191, 200
  feeling of wellbeing 175
First Nation 22, 57, 64, 65
Flavour 38, 40, 51, 172, 174
food 8, 12, 37–8, 40–2 *passim*, 45, 49, 50–1, 66–7, 83–4, 87, 94, 101–3 *passim*, 138, 145, 153, 161
  food miles 179
  food security 180
  foodstuffs 44
  food surplus 48
  gathered food 9
  global food systems 182
  relationships with food 170–5 *passim*, 177, 179
  see also healthy foods; home grown; home grown food vortex; garden, GYO (grow your own); nourishment; traditional food plants ('bush tucker')
forest; forests 14–15, 17–18, 66, 70, 73, 94, 96–8 *passim*, 100, 110, 112, 116–18 *passim*, 120–3 *passim*, 125, 129, 145
  forest-bathing *Shinrin-Yoku* 70
  Glyn Forest (*Coed Gwenllian*) 70
  rainforests 49
  see also deforestation

218 PLANTS MATTER

Framework Convention on Tobacco Control (FCTC) 11, 189–90
fungi 15, 67–8
   fungal networks 15, 145

G

garden 23, 40, 45, 61–2, 68, 110, 161–8 *passim*, 179–82 *passim*, 195
   gardeners 39, 46, 51–2, 161–7 *passim*, 171–6 *passim*, 179–82 *passim*
   gardening 22–3, 45, 161–3, 165–7 *passim*, 169–70, 175, 178–81 *passim*
   Garden Cities 161
   GYO (grow your own) 163–4, 168, 171–2, 174–5, 177–8, 180–2 *passim*
   see also health benefits of gardening
gathering 3, 4, 5, 9
   see also hunter gatherers
grammar 3, 6
Guattari, Félix 110–15 *passim*, 124, 138, 140

H

hair 96–8 *passim*, 203
   hairy 101
   pony tail 96
Hall, Michael 146
healer; healers 21, 60, 68, 73, 110–11, 116, 118, 122, 124, 126–7, 129
   Bantu healer 126, 129
   curandeiros 60
   healer's injury 120
   plants as healers 70
   traditional healers 60, 68
healing 20, 41, 58, 61, 69, 112, 116–18 *passim*, 120–1, 124, 126, 129, 153
   healing the land 69
   healing triptych 129
   healing with plants 111
   healing wounds 41
   spiritual healing 66
health 11, 13, 15, 22, 29, 57, 60, 62, 64–5, 70, 72–3, 128, 150, 161, 165–7 *passim*, 171, 174, 179, 182, 190, 209–10
   healthy 41, 67, 87, 177, 180
   healthier 145, 177
   Department of Health 177
   global health 59, 127
   health and well-being 64, 171
   health benefits of tobacco 11
   health benefits of gardening 23
   health care 59
   healthy foods 41, 67, 177, 180
   mental health 70, 165
   public health 190, 200, 207
   see also ill-health; World Health Organization (WHO)
herbalist 71, 129
herbal tea 152
herb qualities 147, 151, 152
herbal medicine 22, 59, 60–1, 70–1, 147, 149
herbal pharmacopoeia 61
herbal remedies 60
herbivores 13, 19, 38, 148
herbivory 144
Himalayan balsam, *Impatiens glandulifera* 71
home 23, 45, 86, 100, 102, 104, 163–6 *passim*, 182, 191, 205
homes 7, 94–5, 171, 196
   home cooking 174
   home-garden 163–4, 175, 182
   home grown 178, 182–3, 198
   Home Grown Food Vortex 178
   homeland 201
   homelife 175
   homemade 178
   'stay-at-home' order 22, 161, 180
   home of Kagkbusánkua 86
   see also world-house
horticulture 45
Hulme, Keri 195
human nature 124
hunter gatherer 9, 18, 44–5, 47, 169

hunting 9, 181
   'man the hunter' 9

I

ill-health 193
   see also sickness
immersion 23, 83, 143, 167–9 *passim*, 179, 182–3 *passim*
Indigenous 15, 22, 28, 29, 57, 60, 63, 64, 66, 69, 73, 84, 98, 101, 117, 191, 192, 205
   Indigenous Amazonia 110
   Indigenous knowings; knowledge 27, 28, 60, 73, 126
   Indigenous peoples 70, 189
   Indigenous practitioners 11, 13, 27
   Indigenous rights 64
   Indigenous territories 64
   Non-Indigenous 105
infusion 11
ingestion 3, 11, 20, 22, 41–2, 147, 150
Inglewood (Queensland, Australia) 195–7 *passim*, 200, 201, 205
insects 13, 19, 38, 72
   insect behaviour 144
insecticides 206
interdependencies 5, 8, 25, 48
intoxicants 42
investment 38, 42, 46

J

Japanese knotweed, *Reynoutria japonica* Houtt 72

K

kággaba 95–6, 106
kagksouggi 84, 99–103 *passim*, 106
   Kagksouggi People 101–3 *passim*
Kew 67–8, 71
Khoisan 127
kin 15, 17–18
   kincentric ecology 65
   kinship, cross-species 8
   kinship ties 17
kin plants 14, 17
knowledge 10, 28–9, 60–3 *passim*
   ancestral knowledge 29
   ecological knowledge 29
   knowledges 2, 27–8
   plant knowledge 61, 63
   local knowledge 28
   see also TEK, traditional ecological knowledge
Kogi 15–16 *passim*, 28, 83–92 *passim*, 95–106 *passim*

L

language 6, 11, 19, 23–5 *passim*
   language of relationships 23–5 *passim*
   see also grammar; linguistic relativity; plants' languages; Sapir-Whorf hypothesis
Law of the Mother 87
life cycle 44, 47
linguistics 113, 193
   linguistic relativity 25
listening 28, 69, 104, 128

M

material; materials xviiii, xx, 1–6 *passim*, 14, 17, 18, 23, 25, 26, 37, 38, 40, 41, 47, 50, 57, 68, 72, 92, 103, 136–9 *passim*, 141, 147, 148, 150, 154, 163, 197,198, 209
   material culture 45, 48, 52, 53, 189
   material field 10
   material object 195
   material world 24, 48
   see also New Materialities
materialise 150
materialised 83, 92, 93, 96
materialism, vital 109, 146
materalities 110–11, 123
   non-human materialities 114
materiality xviiii, 1, 2, 5, 6, 10, 11, 16–18, 20, 23, 26–7, 37, 42, 48, 50, 52, 118, 121, 126

matters, the grammar of 3
Mbuti 18
medibility 22, 146–7
medicinal plants 9, 57, 58, 60–3
  passim, 66–8, 71–3 passim,
  112, 117, 127, 136, 138, 147–51
  passim, 153–4
  medicinal properties 95
meshworks 37, 47, 48, 52
  meshwork of flows 163
metabolites 148–51 passim
  see also secondary plant metabolites
methyl bromide 206
Mother Earth 95, 105
Mother, the Great 88–91 passim, 96, 104
  Mother of creation 86
  see also Law of the Mother
mother plant 17
mother trees 15–16, 100, 101, 145
  see also trees
Motueka (Aotearoa/New Zealand) 194–200 passim, 204–5
multi-species; multispecies 8, 17, 147, 189
mutual 40, 43, 44, 52, 105, 127
  mutualisms 8, 15
  mutualistic 39, 47
  mutuality 2
mycorrhizal 15, 68, 145
  see also fungi
Myrtleford (Victoria, Australia) 195–8 passim, 200–2 passim, 205, 207

N

nature 6, 7, 40, 51, 65, 69, 90–2 passim, 95, 104–6 passim, 110, 129, 136, 146, 161–3 passim, 165, 168–70 passim, 175
  nature-based solutions 69
  nature/culture dichotomy 40, 65, 162, 170, 179
  nature of 84, 192, 210
  see also human nature

New Materialities xx, 1–7 passim, 26, 147, 189
Newtonian 5
non-human xviiii, 23, 64, 69, 72, 112, 113, 192
  non-human animals 8, 17, 40, 163, 182
non-agential 24
nourish 57, 101, 103, 104
  nourishing 23, 85, 103, 167
  nourishment 101, 106, 143, 161, 177, 182–3
  un-nourishing 105
nutrition 2, 51–2, 66, 177
  nutritional 41, 169
  nutritionists 174, 177

O

ontology 44, 83, 114, 148
open beings 137–41 passim
  see also split beings
opium 42, 52, 58
  see also Poppy *Papaver somniferum* L.
ouroboros 24

P

personhood 65, 100, 103
perspectivism 102, 192
Peru 11, 12
pesticides 209, 210
pharmaceutical 22, 57, 58, 59, 60, 65, 136
photosynthesis 3
  photosynthesise 135
  photosynthesising 2, 8
phytochemicals 38, 136, 148–51 passim
plantabilities 21, 145, 147
  plant agency 51, 53, 69, 135, 152, 169, 189
  plant calling 117, 118, 121, 129
  plantchantments 140
  plant communication 19–20, 69, 139, 143

plantful; plantfulness 21, 23, 140, 142
plant healers 70
plants, kin 14, 16–18, 64–5, 69, 141, 144–5
plants' languages 11
plant medicine as ecology 135, 146, 150
plantness 142
plant-people entanglements, plant-people relations 37, 45
plants as persons 91, 146
plant science 135, 141, 143, 148, 154
plant teachers 10, 27
  see also vegetal teachers
plant thinking 140
  see also mother plant; becoming plant
plantation 49, 189
Poppy (*Papaver somniferum*) 57–8
Poppy goddess 58
potter 164–5
pottering 22, 23, 163–5, 168
praxis 37, 45

Q
quantum 4, 7

R
recycled 198
reproduce 38–40, 43, 48, 52
  reproduced 140
reproduction 39–40, 42, 47, 51, 86, 144
rhizome 40, 111, 138
rhythms (annual rhythms) 44–6
  rhythms (repeated pattern or sound) 109, 117–18, 120, 122, 128
routine (daily) 23, 44, 164, 165, 180

S
Sapir-Whorf hypothesis 25

secondary plant metabolites 68, 146, 148–51 *passim*
  see also metabolites
sedentism 48, 51
seeds 19, 38, 40, 42–3, 71, 85, 92–3, 98, 103, 121, 125, 136, 139, 191, 202
  cosmic seeds 106
  human seeds 84
  seeds as archives 42
sensation 118, 151, 154
sensations 117
sense 15, 19, 20, 149, 151
senses 4, 6, 23, 143, 144, 154, 167–9 *passim*, 172, 179, 182
sense-abilities 8
sensorial 5, 111, 161
  sensorially 23, 24
sickness 57, 62, 65, 90
slaves 49, 51
slavery 193
slave trade 49, 62
spacetimematterings 7
species 3, 13, 15, 20, 21, 39, 40, 58, 60, 66–9 *passim*, 71–3 *passim*, 87, 90, 94, 126, 128–9, 144, 148, 150
  species loneliness 22
  species reduction 105
  interspecies 17, 69
  cross species 15, 17
  see also multi-species
spirit 61, 65, 96, 97, 110, 127
  spirit-beings 94
  spirit/thought 88
  negative spirits 61, 66, 71
split beings 141, 146–8
split critters 137
status, social 51, 101
stimulants 42, 51
stinging nettles, mainly *Urtica dioica* L. 72
stories 5, 19, 70, 72, 87, 175, 209
  Potawatomi stories 23
  tobacco stories 190, 194, 200
storys 28
sugar 42, 45–8 *passim*, 52, 65, 145
  blood sugar 66

sustainability 68, 179, 183
sweetness 39, 52, 153
synanthropes 40

T

taskscape 23, 166, 182
taste 5, 115, 143, 151, 153, 154, 171–9 passim, 183
  tastes 50, 175, 180, 182
  taste of medicine 150
tea 21, 42, 49, 135, 150–2, 154
  tea tasting 152
thing xviiii, xx, 85
  things xviiii, xx, 2, 3 4 6, 12, 23–6 passim, 47, 85, 87, 89–90, 98, 103–6 passim, 113, 115, 118, 125, 147, 152, 163
  thing-power 114
  living things 166, 169
  transformed things 102
tobacco (nicotiana tabacum) 11–13 passim, 27, 42, 49, 65, 189–210 passim
  tobacco control 193
  tobacco cultivation 189, 190, 192, 200, 201–2, 207, 209–10
  tobacco culture 190, 191
  tobacco heritage 195, 209
  see also Framework Convention on Tobacco Control
trade 41, 44, 52, 59, 67, 73, 145
  see also slave trade
tradition 29, 59, 61, 126, 128, 200
  traditions 85, 86, 88, 90, 91, 147
  traditional 59–61, 63, 64, 66, 7, 69, 70–1, 99, 126, 149
  traditional healers 68
  traditional medicine 59, 116, 151, 153
traditional ecological knowledge (TEK) 64
trees 14–17 passim, 28, 38, 46, 70, 73, 83–106 passim, 121, 123, 143, 145
  Kagksouggi trees 84, 100, 103
  Mother trees 15, 16, 145
  Father and Mother of trees 102
  Parents of Trees 103
  see also bark; forest; World Tree
Tree of Knowledge 88, 104, 105
Tree of Life 85, 89
trees as persons 91
triptych 111, 116–18 passim, 120, 129

U

United Nations 190

V

Vegetal 21, 27, 96, 110–17 passim, 120, 122, 125, 127–9 passim
Vegetal beings 8
vegetal democracy 139
vegetal philosophy 21, 135, 148, 149, 154
vegetal teachers 10
vibration 111, 115, 117–20 passim, 144
vitality 21, 97, 110, 121, 123, 127–9 passim
vital materialism 109
VOCs (volatile organic compounds) 13–14

W

warm data 25
weeds 38–41 passim, 44, 72
Wik 63–7 passim
World Health Organization (WHO) 11, 189, 190
world house 95
World Tree, the 85, 86, 88, 94, 103–6 passim
worlding 25, 28, 29, 149, 151

X

Xhosa 110, 126